NONLINEARITY AND CHAOS IN MOLECULAR VIBRATIONS

NONLINEARITY AND CHAOS IN MOLECULAR VIBRATIONS

GUOZHEN WU

Physics Department,
Tsinghua University,
Beijing 100084, China

2005

ELSEVIER

Amsterdam – Boston – Heidelberg – London – New York – Oxford
Paris – San Diego – San Francisco – Singapore – Sydney – Tokyo

ELSEVIER B.V.	ELSEVIER Inc.	ELSEVIER Ltd	ELSEVIER Ltd
Radarweg 29	525 B Street, Suite 1900	The Boulevard, Langford Lane	84 Theobalds Road
P.O. Box 211, 1000 AE Amsterdam	San Diego, CA 92101-4495	Kidlington, Oxford OX5 1GB	London WC1X 8RR
The Netherlands	USA	UK	UK

First edition 2005

Library of Congress Cataloging in Publication Data
A catalog record is available from the Library of Congress.

British Library Cataloguing in Publication Data
A catalogue record is available from the British Library.

ISBN: 0-444-51906-8

♾ The paper used in this publication meets the requirements of ANSI/NISO Z39.48-1992 (Permanence of Paper).
Printed in The Netherlands.

Preface

During my undergraduate and graduate studies, I was very much impressed by group (point) theory. It was then that I learned that many physical laws can be deduced simply from symmetry, not necessarily by Newtonian or quantum mechanics. Hence, symmetry is a more fundamental physical concept than laws in mechanics. Later, I recognized that symmetry breaking is also a core concept in physics. Whereas symmetry can create beauty, its breaking does not necessarily destroy beauty; instead, it may even create another kind of beauty.

Group theory is the study of symmetry and conservation laws in physical systems. A group is often a property of a system only when some interactions are excluded under certain assumptions. In other words, the symmetries of the group are only approximate. However, the situation may not be so simple. The symmetries or the group can be the foundation of the system dynamics. This was the impression I had of this subject in the 1980s.

As I became more familiar with Lie groups, algebras and their applications, I felt a certain disappoinment due to their lack of application in molecular spectroscopy. As I was considering this subject, I saw some publications in this field in international journals, such as those by Professors F. Iachello, R. D. Levine and M. E. Kellman who used Lie group (algebraic) chains and the SU(2) method for an integrable two-body system. I would like to express my gratitude to Professor Kellman for those years when he kindly offered me his copies of preprints and even his unpublished work.

In 1990, while visiting Hong Kong Baptist College (now the University), I started working on the unintegrable three-mode system by the SU(3) group and its algebra. Of course, the work was then new to me and difficult. In turn, through this work, I learnt much of nonlinear dynamics and topics related to chaos, which had become increasingly important since the 1970s. All of these topics were out of my specialty in molecular spectroscopy and chemical physics/physical chemistry and I was startled to find that they could be tractable in my fresh work and they became the core of my concern and interest.

This book is not simply a collection of my works over the past years but intends to summarize the concepts and my concerns. In addition, I have added some background material which is usually scattered in the literature in an unorganized way and is difficult for those who are not very familiar, or are just becoming acquainted, with the field. I hope this addition will help the readers, though it is in no way complete and perfect.

I am reluctant to claim that this approach to highly excited vibration, among the many intelligent algorithms proposed by distinguished scientists around the world in the past ten to twenty years, is ideal and new. The foundation of this approach, the coset representation, is not new, at least in nuclear physics. (For instance, see W. M. Zhang, D. H. Feng and R. Gilmore, *Review of Modern Physics*, 865, vol.62, 1990) We also acknowledge that as an approximation, this approach is equivalent to that by Heisenberg's correspondence. The mainstreams of this approach are:

(1) The Lie group or algebra adopted is not the exact symmetry that a vibrational system possesses. In other words, the system Hamiltonian is not an invariant of the Lie group or algebra. The group or algebra is just the foundation for realizing the vibrational system.

(2) The coset space constructed by Lie group/algebra possesses a geometric structure. Its representations are complete so the system Hamiltonian expressed in the second

quantized operators can be cast in terms of the coset space coordinates. Thus defined, *algebraic* Hamiltonian is physically intuitive and is, hence, easy to write down. Its coefficients are from the fit of its eigenenergies to the experimental data. Then, via the integration of Hamiltonian equations of motion, the motion trajectories can be obtained. The dynamics is described by the trajectories. The coset space is the dynamical phase space. Since highly excited vibrational states are nearly (semi)classical, this algorithm is adequate.

(3) The highly excited vibrational dynamics is nonlinear. Concepts of nonlinear dynamics such as chaos, fractal and Lyapunov exponents will find their roles in such a study. Thus, this algorithm bridges the gap between molecular spectroscopy and nonlinear dynamics. The fruits in the nonlinear field will certainly enrich our understanding of molecular vibration and spectroscopy. In this respect, the vibrational nonlinearity of H_2O, CH_2-, CD_2-, D_2O, H_2S, O_3, the unusual modes of the acetylene C–H bend and their nonergodicity; and the period-3 vibration of DCN as shown in our work are the results of nonlinear dynamics.

(4) This algorithm is also applicable for electronic motion. In this book, its contrast with Hückel molecular orbital theory is stressed. This may deepen our understanding of molecular orbital theory. We will treat only the one-electron system and for two-electronic correlation, the treatment is only superficial. For multi-electron systems, however, this algorithm is possible in principle.

(5) The merits of this algorithm are eminent where the concept of trajectory takes the place of wavefunction. Their contrast, hopefully, may deepen our understanding of quantum mechanics. The idea that wavefunction is the unique paradigm for a quantal or semi-quantal system is not always encouraged. We may expect that the concept of nonlinearity can be beneficial to our understanding of quantum mechanics such as those mentioned in the book: the fractal structure of eigencoefficients, the existence of approximately conserved quantum numbers and the conjecture of quantizing a nonintegrable system by locally minimal Lyapunov exponents, etc. These concepts can be more prevalent than just in the vibrational arena.

Here, we first introduce the concept of dynamical groups. This is the foundation. Then we introduce some concepts of nonlinear dynamics, a very vast and rich area. Here, it is impossible to cover all topics. We just try to emphasize the most useful ones for our study.

The application in molecular vibration and rotation of su(2) algebra, noncompact su(1, 1) algebra, su(3) algebra, h_4 algebra and some nonlinear characters of highly excited molecular vibration are then followed. Furthermore, due to mode nonlinearity and inter-mode couplings, vibrational modes are no longer the normal modes under harmonic approximation as treated in the texts of *traditional* classical dynamics.

Molecular highly excited vibration is a highly nonlinear system. What is its dynamical nature? Though some are mentioned in the book, we have to admit that we still know very little of the dynamics. Then, we introduce some important concepts related to chaos, such as pendulum motion, quasi-periodicity and resonance overlap.

The chapter covering the fractal structure of eigencoefficients indicates that besides probability, they contain also geometric content. This is significant. Traditional quantum mechanics lacks this interpretation. The highly excited bend motion of C–H in acetylene is a hot topic at the moment. What do experimental data tell us about its dynamics? In Chapter 12, we try to answer this.

The concept of the Lyapunov exponent is simple. It shows the divergent deviation rate of two nearby trajectories. The Lyapunov exponent signifies deeply the content of a chaotic system. We have two chapters on its application in molecular vibration and even electronic motion. This is indeed a core topic. Via Lyapunov analysis, we demonstrate that the highly excited C–H bend motion of acetylene may not be ergodic though it is of high energy. It is most important that we propose the conjecture of quantizing a nonintegrable system by locally minimal Lyapunov exponents. The semiclassical quantization of integrable systems by Bohr and Einstein, Brillouin, Keller (EBK) has been perfected. However, for a nonintegrable chaotic system, this is still an open question. Gutzwiller has pointed out that periodic orbits in a chaotic system can play a central role in quantization. We point out that the Lyapunov exponent can also possibly play such a role. D–C and C–N stretches of the DCN molecule constitute a simple system. It is a prototype of coupled molecular vibration. The observation of period-3 trajectories indicates that its dynamics is very complicated, including chaos. We also address the properties of the actions of its periodic orbits and their possible application in system quantization.

The assignment and classification of the highly excited vibrational states seem very difficult, if not impossible, due to complicated inter mode couplings. The fact is that though most conserved quantities are destroyed, some can still be retrieved as approximately conserved quantum numbers and, thereby, the complicated level structures can be assigned and classified. The characteristics of the approximate quantum numbers are also offered.

The chapter on one-electronic motion demonstrates the possible application of this algorithm in the vast arena of physics and chemistry. We try to unify some concepts in diverse fields by such an algorithm. Anderson's localization and the Hammett equation are such attempts.

An analogue to the *H* function by Boltzmann will be derived in Chapter 18 for its application in intramolecular vibrational relaxation and resonance. Dixon dip and its destruction are the topics related to resonance and resonance overlapping which lead to chaos. Their implications will be fully covered in Chapter 19 for the highly excited levels.

Finally, we will touch upon the topic of our current interest: chaos in the transitional and dissociative states. We only mention some of our concerns and ideas. Indeed, they are far from complete. We hope that in the future, more of its nature can be unfolded. My bias is that it bears very essential ideas in physics and chemistry.

Though most of the chapters in the latter parts of the book are self-contained and can be read independently, I still encourage the readers to read them in sequence, so that my ideas, concepts and concerns can be better appreciated.

The work presented in the book is a summary of my work and a record of my labors in this field up to this time. It is far from complete and may even contain errors. The questions it raises may be more than the answers it offers. Indeed, more questions and topics are waiting to be explored and answered. As more experimental data on highly excited molecular vibration become available due to technological achievements, more ideas about its nature are being proposed.

These past years, when I was deeply involved in this challenging and attractive field, were rewarding and full with excitement. The publication of this book is the harvest of all my previous labour. Up until 1999 I worked mostly alone and felt both failure and success, though my career was full with excitement. In the spring of 1999, I visited the Physics Department of the Central University in Taiwan. It seems that the calm campus and the excitement of returning to Taiwan, where I spent my formative years, were a stimulus and some works were finished that time. After 1999, my students joined the work and helped give it depth. Their contributions are, of course, refelected in this book. They are X.D. Ding, Z.Q. Ji, J. Yu, X. Guan and D.S. Zheng, P.J. Wang, and H.R. Wang. Their works, albeit sometimes confirming my predictions and proposals, more often pointed out my errors. In the past years, I have also visited the Center for Advanced Study at Tsinghua University. All these deserve my acknowledgements. The grants by National Natural Science Foundation of China during 1995–1997, 2000–2002 and 2004–2006 are deeply appreciated.

Finally, I am grateful for the opportunity presented by Elsevier in publishing this book, which will help to distribute my work to a larger audience.

Guozhen Wu
wgz-dmp@tsinghua.edu.cn
Spring 2005,
Tsinghua University,
Beijing, CHINA

Contents

Nonlinearity and Chaos in Molecular Vibrations

Chapter 1

Molecular vibration

1.1 Normal modes

Molecules are made of atoms which are connected by chemical bonds. The length of the chemical bonds or the distance between atoms is not fixed. Hence, as a whole, molecules vibrate. Molecular vibration can be considered to be like the motion of particles connected by springs.

Consider a molecule consisting of N atoms with mass m_α (α is the numbering of the atoms) . The displacements of atoms are $\Delta x_\alpha = x_\alpha - x_\alpha^e$, $\Delta y_\alpha = y_\alpha - y_\alpha^e$, $\Delta z_\alpha = z_\alpha - z_\alpha^e$. Here, (x_α, y_α, z_α) are the instantaneous positions of atoms and (x_α^e, y_α^e, z_α^e) are their positions at equilibrium. Therefore, the kinetic energy of a molecule can be expressed as

$$T = \frac{1}{2}\sum m_\alpha [(\frac{d\Delta x_\alpha}{dt})^2 + (\frac{d\Delta y_\alpha}{dt})^2 + (\frac{d\Delta z_\alpha}{dt})^2]$$

To simplify this expression, we define a set of coordinates, q_i as

$$q_1 = \sqrt{m_1}\Delta x_1,\ q_2 = \sqrt{m_1}\Delta y_1,\ q_3 = \sqrt{m_1}\Delta z_1,\ q_4 = \sqrt{m_2}\Delta x_2, \ldots$$

then T can be written as:

$$T = \frac{1}{2}\sum_{i=1}^{3N}\dot{q}_i^2$$

The potential energy, V, among the atoms is the function of q_i, i.e., $V = V(q_1,\ldots,q_{3N})$. We may expand V in terms of q_i:

$$V = V(0,\cdots,0) + \sum(\partial V/\partial q_i)_0 q_i + \frac{1}{2}\sum(\partial^2 V/\partial q_i \partial q_j)_0 q_i q_j + \cdots$$

where $V(0,\ldots,0)$ is the potential energy at equilibrium and can be set to 0. $(\partial V/\partial q_i)_0$ is the value at equilibrium and is also 0. Hence, up to the second order, i.e., under the harmonic approximation, we have

$$V = \frac{1}{2}\sum f_{ij}q_iq_j$$

where $f_{ij} = (\partial^2 V / \partial q_i \partial q_j)_0$ can be considered as the force constant between the two coordinates q_i and q_j.

Now, T and V are the functions of $\dot{q}_i$ and q_i. The Lagrangian $L = T\text{-}V$ is also a function of $\dot{q}_i$ and q_i. From the equation of motion for L :

$$\frac{d}{dt}\left(\frac{\partial L}{\partial \dot{q}_i}\right) - \frac{\partial L}{\partial q_i} = 0\,, \quad i = 1,\cdots,3N$$

we have

$$\ddot{q}_i + \sum_{j=1}^{3N} f_{ij}q_i = 0$$

Suppose that $q_i = q_i^0 \cos(\omega t + \varepsilon)$ and substitute it into the above equations, we get:

$$\sum_{j=1}^{3N}(f_{ij} - \delta_{ij}\omega^2)q_i^0 = 0\,, \qquad i = 1,\cdots,3N$$

($\delta_{ij} = 0$ as $i \neq j$; $\delta_{ij} = 1$ as $i = j$)

For this set of equations to have nonzero solutions, the determinant $|\, f_{ij} - \delta_{ij}\omega^2 \,|$ has to be 0. The number of solutions is 3*N*. For a nonlinear molecule, there are six solutions with $\omega_k^2 = 0$, corresponding to the three translations and three rotations of the molecule as a whole. For a linear molecule, there are two rotations and, hence, five solutions with $\omega_k^2 = 0$.

To a solution with eigenvalue ω_k^2, there corresponds an eigenvector $(q_{1,k}^0, \cdots, q_{3N,k}^0)$. Usually, a normalization condition is chosen such that:

$$\sum q_{i,k}^{0\;2} = 1$$

The normalized $q_{i,k}^0$ is expressed as L_{ik}.

The analysis above shows that in a mode with ω_k, each atom vibrates with the same ω_k while with different amplitude L_{ik}. In a mode, the phase difference between two vibrating atoms can only be 0 (in phase) or π (out of phase) depending on whether their L_{ik}'s are of the same or opposite signs. Such modes are called the normal modes.

Here, we note that, in a normal mode, the phase differences among the atomic displacements are 0 or π, which is the consequence of the harmonic approximation. (For instance, in O_3, there are the symmetric and antisymmetric normal modes of the two O–O stretching motions.) Under the harmonic approximation, we have the eigen routine to elucidate the eigenvalues and eigenvectors. (This is the **GF** matrix algorithm in Ref. 1.1). The eigenvalues correspond to the frequencies and the eigenvectors to the atomic displacements. The eigenvectors are real and the phase differences among the atomic displacements can only be 0 or π depending on whether they are of the same or opposite signs. When *anharmonic* approximation is considered, there will be no routines leading to the elucidation of eigenvalues and eigenvectors. The phase differences among the atomic displacements will not be only 0 or π. Then, there are no concepts such as normal mode and symmetry, antisymmetry as in the harmonic approximation. In Chapter 4, it will be shown that vibrational modes are in fact quite complicated, depending on the energy, the coupling strengths among the bonds and the constants of motion they possess. But, in general, we can say that in the lower states, say, of the two stretches of O_3, the phase difference can be arbitrary, the local character being due to weak coupling. In the higher states the phase will be locked around 0 or π, leading to the symmetric and antisymmetric modes due to stronger coupling. Furthermore, due to nonlinearity, localized modes may show up in the higher states. These will be the topics of the subsequent chapters.

For a normal mode with ω_k, there is the normal coordinate Q_k. Q_k is related to q_i as:

$$q_i = \sum_k L_{ik} Q_k$$

or

$$Q_k = \sum_i L_{ik} q_i$$

With Q_k and $\dot{Q}_k$, $T = \frac{1}{2}\sum \dot{Q}_k^{\,2}$, $V = \frac{1}{2}\sum \omega_k^{\,2} Q_k^{\,2}$ and the equations of motion are:

$$\ddot{Q}_k + \omega_k^{\,2} Q_k = 0, \quad k = 1, \cdots, 3N-6$$

This is a simple harmonic motion. (For its detailed description, see Refs 1.1.)

We recognize that the concept of normal mode is based on the harmonic approximation. Under harmonic approximation, there are no couplings among the normal modes. The normal mode is a simple harmonic oscillator.

Therefore, the vibrational energy of a molecule is

$$\sum_k \hbar\omega_k (n_k + \frac{1}{2})$$

with n_k the quantum numbers.

In fact, as the vibrational energy is higher, the harmonic approximation is no longer valid, and neither is the concept of normal mode. Sticking to the concept of normal mode, we can still consider the nonlinear effect in terms of the couplings among the normal coordinates. This is a perturbational treatment. However, this treatment is limited since the motion is nonlinear and its essence is different from that of linearity, even modified with perturbation.

1.2 Morse oscillator

The concept of normal mode can indeed explain many phenomena in molecular spectroscopy. However, it has an inherent deficiency in that it cannot accommodate the dissociation (breaking) of chemical bonds at high excitation. The simplest model that can accommodate dissociation is the Morse oscillator.

Morse potential is

$$D\{1 - \exp[-a(r - r_0)]\}^2$$

Here, a, D are constants, r_0 is the equilibrium position, $r - r_0$ ($\equiv \Delta r$) is the displacement and a^{-1} is the characteristic length. Figure1.1 shows the relation of the potential against Δr where the horizontal line is the level with action n. Δr_0^+, Δr_0^- show its extrema of displacement.

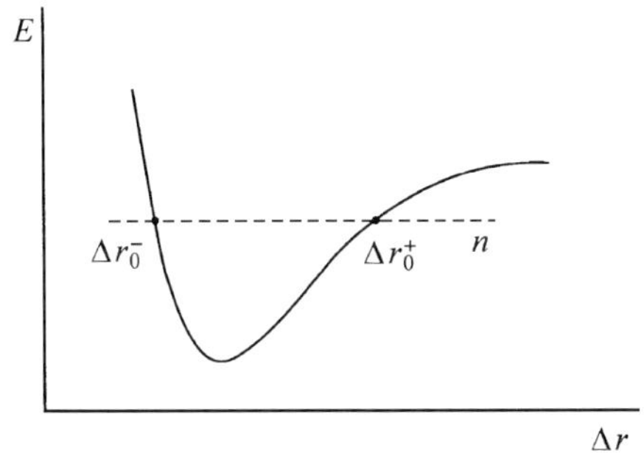

Fig.1.1 The Morse potential. Δr_0^+,Δr_0^- show the extrema of displacement for the level (the horizontal line) with action n.

The action n of a Morse oscillator can be written as $(1/2)(q^2 + p^2)$. (q, p) are the generalized coordinates. For convenience, we can consider q as the quantity related to the displacement Δr and p the quantity related to the momentum of the oscillator.

For the elucidation of the dynamical properties of the Morse oscillator, we can start from the solution of its Schroedinger equation. However, we will try the other way. Suppose as $p = 0$, the displacement reaches its extrema, denoted as Δr_0^+ and Δr_0^-. Their relations with q as shown below will lead to the expression of D， k（See below for its definition）of the Morse oscillator in terms of the spectroscopic parameters and *vice versa*. The whole procedure, though lacking in rigor, is very beneficial to our understanding of the characteristics of the Morse oscillator.

The relation between Δr_0 and q is (See Ref.1.2)

$$\Delta r_0 = a^{-1}\ln\{[1- (1-\lambda^2)^{1/2}\text{sgn}]/ \lambda^2\} \qquad (1.1)$$

Here, as $q < 0$, sgn = 1, Δr_0 is Δr_0^- ; as $q > 0$, sgn = -1, Δr_0 is Δr_0^+. Meanwhile, $\lambda = 1-(2n+1)/k$. k is a parameter. Its physical meaning will be given later. As $p = 0$， we have $2n = q^2$.

The relation between $\Delta r_0 \times a$ and q is shown in Fig. 1.2.

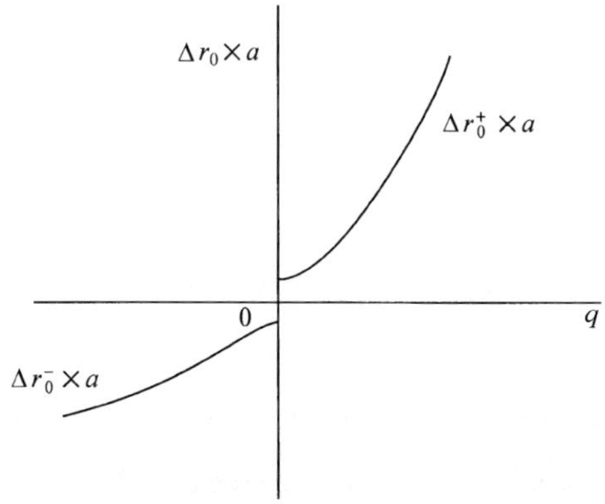

Fig.1.2 The relation between the displacement $\Delta r_0 \times a$ and the generalized coordinate q of the Morse oscillator.

From (1.1), we have:

$$1-\exp(-a\Delta r_0)= -(1-\lambda^2)^{1/2} \quad \text{sgn} =1,$$
$$(1-\lambda^2)^{1/2} \quad \text{sgn} = -1$$

or

$$[1-\exp(-a\Delta r_0)]^2=1-\lambda^2$$

This means that the Morse potential is: $D(1-\lambda^2)$

Now, let's consider the state with action n.

Since $\lambda=1-(2n+1)/k$

then,

$$D(1+\lambda)(1-\lambda) = (4D/k)(n+1/2)-(4D/k^2)(n+1/2)^2$$

Note that $D(1+\lambda)(1-\lambda) = D(1-\lambda^2)$ is the potential at the extremal displacements. At these displacements, the kinetic energy is zero. Hence, $D(1-\lambda^2)$ is the state energy with action n. Under quantization, n is 0, 1, 2⋯, the quantum numbers and $D(1-\lambda^2)$ is the eigenenergy. The right-hand-side expression shows that this is a second order anharmonic oscillator with $-4D/k^2$ the second order nonlinear coefficient and $4D/k$ the linear coefficient which is related to the frequency of the harmonic motion. Therefore, the Morse oscillator is a system of second order anharmonicity. In spectroscopy, we generally denote the eigenenergy E_n of such a system as

$$\omega(n+1/2)+X(n+1/2)^2 \qquad (1.2)$$

Hence, for a Morse oscillator:

$$\omega = 4D/k, \quad X = -4D/k^2$$

or

$$k = \omega/(-X), \; D = (\omega/2)^2/(-X)$$

The significance is that we now relate the Morse parameters k, D to the spectroscopic parameters ω, X.

The energy spacing between two nearest neighboring levels becomes smaller as n is larger. As the spacing is zero, the oscillator dissociates. We denote the quantum number as n_0 at dissociation. From

$$E_{n_0+1} - E_{n_0} = 0$$

we have:

$$\omega + X(2n_0+2)=0$$

or

$$n_0 = (1/2)(\omega/(-X))-1 = (k/2)-1 \approx k/2$$

Hence, k is twice the dissociation quantum number which can be obtained from the spectroscopic parameters ω and X.

Finally, we note that $a = \omega\sqrt{\mu/2D} = \sqrt{-2X\mu}$ with μ the reduced mass of the Morse oscillator.

Substituting $n_0 = (1/2)(\omega/(-X))-1$ into (1.2), we can readily calculate the dissociation energy as $(\omega^2-X^2)/(-4X) \sim D$. For instance, for the D---CN bond, from experiment we have $\omega = 2681.4$ cm^{-1}, $X = -21.2$ cm^{-1}, then $n_0 \sim 63$. In reality, the dissociation quantum number is less. This is due to the inter-mode coupling which is prone to dissociation.

1.3 Second quantized operators

The canonical form of the Hamiltonian of a harmonic oscillator is

$$H = \frac{1}{2}(p^2 + \omega^2 q^2)$$

with p, q the conjugate momentum and coordinate. Quantization requires that they obey the following commutation relation:

$$[q, p] \equiv qp - pq = i\hbar$$

Hence, we have

$$[H, p] = i\hbar\omega^2 q \text{ and } [H, q] = -i\hbar p$$

Let $|n\rangle$ be the eigenfunction of H, i.e., $H|n\rangle = E_n|n\rangle$ and define $a^+ = -ip + \omega q$

Then we have:

$$\begin{aligned} Ha^+|n\rangle &= H(-ip + \omega q)|n\rangle \\ &= \{-i[H, p] - ipH + \omega[H, q] + \omega qH\}|n\rangle \\ &= \{-i(i\hbar\omega^2 q) - ipE_n + \omega(-i\hbar p) + \omega qE_n\}|n\rangle \\ &= (E_n + \hbar\omega)a^+|n\rangle \end{aligned}$$

Therefore, $a^+|n\rangle$ possesses more energy of an amount of $\hbar\omega$ than that of $|n\rangle$.

Similarly, define $a = ip + \omega q$, then

$$Ha|n\rangle = (E_n - \hbar\omega)a|n\rangle$$

Therefore, $a|n\rangle$ possesses less energy of an amount of $\hbar\omega$ than that of $|n\rangle$.

We can apply the a operator several times for the oscillator to reach its ground state $|0\rangle$, then

$$a|0\rangle = 0$$

From

$$\begin{aligned} a^+a|0\rangle &= (p^2 + \omega^2 q^2 + i\omega[q, p])|0\rangle \\ &= (2H - \hbar\omega)|0\rangle \\ &= (2E_0 - \hbar\omega)|0\rangle = 0 \end{aligned}$$

hence,

$$E_0 = \frac{1}{2}\hbar\omega$$

The analysis above shows that the ground energy of the oscillator is $\hbar\omega/2$ and that a^+ excites the oscillator with an amount of energy $\hbar\omega$. In other words, the energy spectrum is:

$$E_n = (n+\frac{1}{2})\hbar\omega\,, \quad n = 0, 1, 2, \ldots,$$

and

$$H|n\rangle = (n+\frac{1}{2})\hbar\omega|n\rangle$$

$$a^+|n\rangle \sim |n+1\rangle$$

$$a|n\rangle \sim |n-1\rangle$$

Specifically, if we require that

$$a^+|n\rangle = \sqrt{n+1}\,|n+1\rangle$$

$$a|n\rangle = \sqrt{n}\,|n-1\rangle$$

then a^+ and a can be written as

$$a^+ = A(-ip+\omega q)\,, \quad a = A(ip+\omega q)$$

A can be fixed by the following consideration:

$$\begin{aligned} aa^+|n\rangle &= (n+1)|n\rangle \\ &= A^2(ip+\omega q)(-ip+\omega q)|n\rangle \\ &= A^2\{2H + i\omega[p,q]\}|n\rangle \\ &= A^2 2(n+1)\hbar\omega|n\rangle \end{aligned}$$

i.e.,

$$A = 1/\sqrt{2\hbar\omega}$$

Hence, we call $a^+ = \frac{1}{\sqrt{2\hbar\omega}}(-ip+\omega q)$ the creation operator and

$a = \frac{1}{\sqrt{2\hbar\omega}}(ip+\omega q)$ the destruction operator.

By these two operators, we have:

$$H = \hbar\omega(a^+ a + \frac{1}{2})$$

$$[a, a^+] = 1, \ [a^+, a^+] = [a, a] = 0$$

and

$$a^+ a|n\rangle = n|n\rangle$$

$a^+ a$ or n is the number operator. All these are very useful formulae. Of course, we can represent p, q in terms of a^+ and a as:

$$p = i\sqrt{\frac{\hbar\omega}{2}}(a^+ - a)$$

$$q = \sqrt{\frac{\hbar}{2\omega}}(a^+ + a)$$

For a Morse oscillator, the relations between (q, p) and (a^+, a) are not as straightforward as for the harmonic oscillator. However, we will still adopt most of those formulae of the harmonic oscillator as shown in this section for the Morse oscillator. For instance, for the Morse oscillator, we still have:

$$a^+|n\rangle = \sqrt{n+1}\,|n+1\rangle$$

$$a|n\rangle = \sqrt{n}\,|n-1\rangle$$

This is true for small anharmonicity. This formulation can be viewed as a *definition* and will be convenient for our vibrational treatment such as in the elucidation of the coefficients of the algebraic Hamiltonian by fitting to the experimental data as will be shown in the next section.

For the fermions, like electron, we have analogously the creation and destruction operators: b^+ and b. They obey the anticommutation relation:

$$[b, b^+]_+ \equiv bb^+ + b^+ b = 1,$$

$$[b^+, b^+]_+ = [b, b]_+ = 0$$

When we denote the vacuum state as $|0\rangle$, then $b^+|0\rangle = |1\rangle$, showing that there is one electron in a state. Since

$$b^+b^+|0\rangle = \frac{1}{2}[b^+, b^+]_+|0\rangle = 0$$

this means that two electrons cannot share a common state, simultaneously. Moreover, for two states, s and t, since

$$b_s^+ b_t^+ = -b_t^+ b_s^+$$

the interchange of the subscripts s and t is antisymmetric. Apparently, anticommutation guarantees Pauli's exclusion principle.

We also have the number operator for an electron : $b^+b = n$. Since

$$n^2 = b^+bb^+b = b^+(1 - b^+b)b = b^+b = n$$

that is $n = 0$ or 1. This indicates that in a state there can only be one electron at most.

1.4 Algebraic Hamiltonians

For the system of two vibrational modes, s and t, we can consider it as two coupled Morse oscillators. Its Hamiltonian consists of the terms:

$$\omega_s(n_s + \frac{1}{2}) + X_s(n_s + \frac{1}{2})^2 + \omega_t(n_t + \frac{1}{2}) + X_t(n_t + \frac{1}{2})^2$$
$$+ X_{st}(n_s + \frac{1}{2})(n_t + \frac{1}{2})$$

For convenience, n_s and n_t are chosen dimensionless and ω_s, ω_t, X_s, X_t, X_{st} are in the unit of cm^{-1}. (Note we will no longer write $\hbar\omega$.)

Furthermore, we consider that there are resonances (couplings) between the s and t modes. The corresponding expressions are:

$$K_{st}(a_s^+ a_t + a_t^+ a_s)$$

and

$$K_{DD}(a_s^+ a_s^+ a_t a_t + a_t^+ a_t^+ a_s a_s)$$

The former shows the 1:1 resonance while the latter the 2:2 resonance, which is also called the Darling–Dennison resonance.

For the Hamiltonian without couplings, n_s and n_t are conserved. For the 1:1, 2:2 resonances, n_s and n_t are not conserved. However, $n_s + n_t$ is still conserved. This is intuitive if we recognize the physical interpretation of creation and destruction operators.

For the system of three modes, like the system of H_2O with two O–H stretches and one bend, its Hamiltonian can easily be written down. Now, there can be the resonance term of the type:

$$K_{sbb}(a_s^+ a_b a_b + a_t^+ a_b a_b + h.c.)$$

This is the Fermi resonance. Here, $h.c.$ stands for Hermitian conjugate. Therein, n_s, n_t and n_b are no longer conserved. However, $n_s + n_t + n_b/2$ is still conserved. These conserved quantum numbers are usually called the *polyad numbers.*

That polyad number is a constant of motion can be shown by its commutation with the Hamiltonian operator. For instance, for 1:1 resonance, we have:

$$[H,(n_s + n_t)] = [(a_s^+ a_t + a_t^+ a_s),(n_s + n_t)] = 0$$

by noting that $[a_\alpha^+, n_\alpha] = -a_\alpha^+$ and $[a_\alpha, n_\alpha] = a_\alpha$.

The eigenstates of the algebraic Hamiltonian are easily elucidated. For the system of two modes, since $N = n_s + n_t$ is conserved, we have $N+1$ basis states denoted as $\{|n_s, n_t\rangle\}$ which are: $|0,N\rangle$, $|1,N-1\rangle$, …, $|N,0\rangle$ with N given. Then by these $N+1$ basis states, an Hamiltonian matrix with elements: $\langle n_s', n_t'|H|n_s, n_t\rangle$ can be constructed. By diagonalizing the matrix, the eigenstates and their energies are obtainable. Of course, in this process, the coefficients of the algebraic Hamiltonian have to be known in advance. So, in practice we need a trial to determine the coefficients by the fit of the calculated eigenenergies to the spectroscopic data using a nonlinear algorithm such as the Marquardt method. In this sense, the algebraic Hamiltonian does possess the content from the experiment and is not a purely theoretical supposition! In fact, the resonance types of a system are determined by this way of fitting.

Later on, we will consider the electronic motion in the system of multi-sites. Its algebraic Hamiltonian can be written as

$$\sum_K \varepsilon_k a_k^+ a_k + \sum V_{ij} a_i^+ a_j$$

Here, ε_k is the energy at site k. V_{ij} is the exchange energy between sites i and j.

References

1.1. E. B. Wilson, J. C. Decius and P.C. Cross, *Molecular Vibrations*, McGraw-Hill, New York, 1955

1.2. C.C. Rankin and W.H. Miller, *J. Chem. Phys.* 55 (1971) 315

Chapter 2

Concepts of dynamical groups

2.1 Continuous groups

A group is an important concept in mathematics. Its definition is simple. It is a collection of elements with an operation * that satisfy the following relations:

(1) $a * b = c$ (a, b, c are elements. c is unique.)

(2) There is an element called identity e, such that for a, there is b, and $a * b = b * a = e$. b is called the inverse of a and is denoted as a^{-1}.

(3) There is the associative property: $(a * b) * c = a * (b * c)$

The number of the elements can be finite, infinite, discrete or continuous.

The geometric shape of a molecule may possess symmetries. The symmetry elements can form a point group. Most often, this will be finite and discrete. Point groups are useful in the analysis of molecular vibration. In the realm of molecular highly-excited vibration, we are concerned with continuous groups.

The elements of a continuous group can be parameterized. For instance, the rotations along an axis form a continuous group. Its elements can be represented by the corresponding rotational angle θ. The rotations in three dimensional space also form a continuous group. Each of its elements $\hat{O}$ can be represented by the corresponding Euler angle. That is, we have $\hat{O}(\theta,\phi,\chi)$. In general, we will have a parameter space to represent the elements of a continuous group. Its operation is to associate two points with a third point in the parameter space. Hence, this parameter space possesses not only topological (geometric) properties but also the algebraic (operational) properties. Continuous groups are therefore called topological groups. (Here, we do not stick to the rigorous mathematical definition.) The coexistence of topological and algebraic properties is a very important characteristic. This means that the group properties can show up both topologically and algebraically. The topological space of the group of rotations along an axis is equivalent to the real number axis. The parameter space for the rotational group in the three dimensional space is a two dimensional unit sphere. Suppose that in the parameter space we can choose an adequate coordinate system such that its points (group elements) α and β are represented as $(\alpha^1, \alpha^2 \cdots \alpha^\eta)$ and $(\beta^1, \beta^2 \cdots \beta^\eta)$. If $\gamma = \alpha * \beta$, then there are functions ϕ^μ such that

$$\gamma^\mu = \phi^\mu(\alpha^1, \alpha^2, \cdots \alpha^\eta, \beta^1, \beta^2, \cdots \beta^\eta), \quad \mu = 1,2,\cdots \eta$$

If ϕ^μ is differentiable with respect to α^i and β^j, then the continuous group is also called a Lie group.

Near the identity, which can be represented as (0, 0, …, 0), we can construct the operators:

$$X_\lambda = -\sum_\mu \frac{\partial \phi^\mu(\beta,\alpha)}{\partial \beta^\lambda}\bigg|_{\beta=0} \frac{\partial}{\partial \alpha^\mu}$$

It can be shown that these operators satisfy the following relations:

$$[X_\mu, X_\nu] = c^\lambda_{\mu\nu} X_\lambda \text{ (sum over } \lambda \text{, for repetitive index)}$$

$$[\lambda_1 X_\mu + \lambda_2 X_\nu, X_\lambda] = \lambda_1[X_\mu, X_\lambda] + \lambda_2[X_\nu, X_\lambda]$$

$$[X_\lambda, \lambda_1 X_\mu + \lambda_2 X_\nu] = \lambda_1[X_\lambda, X_\mu] + \lambda_2[X_\lambda, X_\nu]$$

and

$$[X_\mu, X_\nu] = -[X_\nu, X_\mu]$$

$$[X_\mu,[X_\nu, X_\lambda]] + [X_\nu,[X_\lambda, X_\mu]] + [X_\lambda,[X_\mu, X_\nu]] = 0$$

Here, [,] is called the Lie bracket. It is defined as

$$[X_\mu, X_\nu] = X_\mu X_\nu - X_\nu X_\mu$$

$c^\lambda_{\mu\nu}$ are the structural constants. They show the characteristics near the group identity.

The collection of $\{X_\lambda\}$ is, in fact, a vector space. It differs from the common vector space in that it additionally possesses the properties by the Lie bracket. It is called a Lie algebra. A more formal way of describing the above statement is that the tangent space around the identity is a vector space and that its elements (vectors) possess a Lie algebraic structure.

The importance of a Lie algebra is that the elements of the Lie group can be associated with the vectors of the Lie algebra such that:

$$\alpha^{\mu} X_{\mu} \text{ (vector of the Lie algebra)} \rightarrow \exp(-\alpha^{\mu} X_{\mu}) \text{ (element of the Lie group)}$$

(Strictly speaking, this is true only for a simply connected Lie group for which there are no 'holes' in its topological space.)

Lie algebra shows the structure near the identity of the Lie group. This structure can be transferred to those around the other elements. This is because $a \cdot e = a$.

Generally speaking, different Lie groups possess different Lie algebras. But this is not always the case. Two different Lie groups may possess the same Lie algebra. In other words, given a Lie algebra, we are not definitely sure what its corresponding group is. Corresponding to a Lie algebra, there is definitely a unique, simply connected group. Other groups sharing the given Lie algebra can be derived from this simply connected group.

The Lie group is closely associated with Lie algebra. This shows the close relation between topology and algebra. The theory concerning the Lie group and Lie algebra is beautiful. It is strongly recommended that readers consult other treatises for this fascinating field.

Customarily, upper case alphabets are used for the Lie group and lower case alphabets for the Lie algebra.

2.2 Coset spaces

If a subset of a group forms a group, it is called the subgroup of the main group. For a group G and its subgroup H, we can form the set gH (or Hg) with $g \in G$. The set gH (or Hg) is called the left (right) coset. Two cosets are identical when they share a common element. Otherwise, they share no elements in common. Hence, a group can be decomposed as a sum of 'non-intersecting' cosets. If each coset is considered as an element, then we have the coset space. The coset space of a finite group is plain. It consists only of finite elements, whereas for a continuous group the coset space will possess topological (geometric) structure. The symbol *G/H* shows the coset space of *G* with respect to its subgroup *H*.

The rotations in three dimensional space form a Lie group. It is represented as SO(3). Its Lie algebra is so(3). It is formed by the well known angular momentum operators $\{J_x, J_y, J_z\}$ or $\{J_z, J_+, J_-\}$ with $J_+=J_x+iJ_y$, $J_-=J_x-iJ_y$. The group SU(2) of two dimensional unitary matrices with determinant 1 (unitary matrix U satisfies $U^+=U^{-1}$. U^+ is the transpose conjugate and U^{-1} is the inverse) also possesses the same Lie algebra for which $\{J_z, J_+, J_-\}$ are generally chosen as its generators and is denoted as su(2). In group language, we say that SO(3) and SU(2) are homomorphic.

The set $\{J_z\}$ forms a subalgebra of su(2) and is denoted as u(1). The group formed thereby is U(1) with elements e^{cJ_z} (*c* is complex number). The coset space is SU(2)/U(1), whose elements are formed by J_+, J_- and can be written as $\Omega=\exp(\tau J_+ - \tau^* J_-)$.

The eigenstates $|J,J_z\rangle$ of J_z (with *J* fixed) form the bases of an irreducible representation of su(2). Various J's show different irreducible representations. J_z shows the weight for a J irreducible representation. We can choose $|J,-J\rangle$ (or $|J,J\rangle$), i.e., the state of the lowest (or highest) weight, to form the coset representatives $|J\Omega\rangle$

$$|J\Omega\rangle \equiv \Omega\,|J,-J\rangle = \exp(\tau J_+ - \tau^* J_-)\,|J,-J\rangle$$

Note that $|J\Omega\rangle$ and the elements of SU(2)/ U(1) are in one-to-one correspondence. $\{|J\Omega\rangle\}$ (various τ for various coset elements) is overcomplete. This means that for an operator $\hat{O}$, we need only the diagonal terms $\langle J\Omega|\,\hat{O}\,|J\Omega\rangle$ to fully represent its dynamical properties.

Since the coset space SU(2)/ U(1) is a two dimensional sphere (note that J is fixed and the space is formed by J_+ and J_-), τ can be written as the function of (θ,ϕ). For instance,

$\tau = \tan\frac{\theta}{2} e^{-i\phi}$ ($0 \le \theta \le \pi$, $0 \le \phi \le 2\pi$) or other appropriate functional forms.

This parameterization of τ is crucial for the later treatment of molecular problems. It is an important step in the application of the coset approach to physical problems.

The above concept can be generalized as follows:

(1) For a Lie group *G*, choose its subgroup *H* and construct the coset space *G/H*.

(2) To obtain the corresponding Lie algebras *g* and *h*.

(3) For an irreducible representation Γ_Λ of the Lie algebra, choose the state of the highest (or the lowest) weight $|\Lambda,\Lambda\rangle$ ($|\Lambda,-\Lambda\rangle$) as its reference state.

(4) $|\Lambda,\Lambda\rangle$ is the eigenstate of h_α of *h*, i.e., $h_\alpha|\Lambda,\Lambda\rangle = \phi_\alpha|\Lambda,\Lambda\rangle$. Or that for H_α of *H*, we have $H_\alpha|\Lambda,\Lambda\rangle = \exp(i\phi_\alpha)|\Lambda,\Lambda\rangle$.

(5) The coset representatives are

$$|\Lambda,\Omega\rangle = \exp(\sum \tau_\beta E_\beta - h.c.)|\Lambda,\Lambda\rangle$$

where τ_β is the parameter (usually the complex number). E_β is the element of *g* that cannot destruct $|\Lambda,\Lambda\rangle$ (E_β, E_β^+ are the elements of *g* besides *h*)

(6) For an operator $\hat{O}$, we have the representation $\langle\Lambda,\Omega|\hat{O}|\Lambda,\Omega\rangle$.

(7) *G/H* possesses a topological (geometric) structure which can be parameterized by τ_β as its coordinates.

2.3 Dynamic applications

The core concept of the classical mechanics is the Poisson bracket { , }. In the generalized coordinates (q, p), the Poisson bracket of two functions $f(q, p)$ and $g(q, p)$ is

$$\{f,g\}=\frac{\partial f}{\partial q}\frac{\partial g}{\partial p}-\frac{\partial g}{\partial q}\frac{\partial f}{\partial p}$$

For a dynamic quantity $A(q,p)$, its equation of motion is

$$\frac{d}{dt}A(q,p)=\{A(q,p),H(q,p)\}$$

This equation is equivalent to the following Hamilton's equations:

$$\frac{\partial H(q,p)}{\partial p}=\frac{dq}{dt},\quad \frac{\partial H(q,p)}{\partial q}=-\frac{dp}{dt}$$

Here, $H(q,p)$ is the Hamiltonian of the system.

For Heisenberg's equation of motion in quantum mechanics:

$$\frac{dA}{dt}=-i[A,H]$$

We can use $|\Lambda\Omega\rangle$ to represent it, i.e., $\frac{d}{dt}\langle\Lambda\Omega|A|\Lambda\Omega\rangle=-i\langle\Lambda\Omega|[A,H]|\Lambda\Omega\rangle$, with the appropriate choice of generalized coordinates (q,p) for the coset space. At this point, if

$$\langle\Lambda\Omega|[A,H]|\Lambda\Omega\rangle=i\{A(q,p),H(q,p)\},$$

then $$-i\langle\Lambda\Omega|[A,H]|\Lambda\Omega\rangle\Big|_{QM}=\frac{dA(q,p)}{dt}\bigg|_{QM}=\frac{dA(q,p)}{dt}\bigg|_{CLASSICAL}$$

In other words, the dynamic behavior of a quantum system can be exactly described by the Hamilton's equations.

It can be shown that only if A, H are the linear functions of the elements of the Lie algebra, the above equation will hold. Hence, in general, the Hamilton's equations in the coset space are only the approximate description of the quantal behavior of the system. We have to note that here we are not solving the classical behavior as the Planck constant is approaching zero. Instead, we are solving the quantal or semiquantal behavior. The Hamilton's equations in the coset space still possess the quantal character of the system though they show only an approximation if H and A are not the linear functions of the elements of the Lie algebra. This algorithm is built on the mean field approximation based on quantum mechanics. Coset space is a quantum phase space. Hence, we are studying the quantal issue by the classical algorithm. It is the classical analogy for the quantal issue and not the classical limit (as the Planck constant is zero).

With the above mathematical tool, we are ready to explore the molecular vibrational and electronic behaviors.

2.4 Algebraic difference of molecular vibrational and electronic properties

The quantized molecular vibration is a boson system while that of electrons is a fermion system. Their creation (a^+) and destruction (a) operators satisfy different algebraic properties. Bosons satisfy the commutation ([,]) while fermions satisfy anticommutation $[\ ,\ ]_+$ ($[A, B]_+ = AB+BA$). That is, for bosons:

$$[a_i, a_j^+] = \delta_{ij},\quad [a_i, a_j] = [a_i^+, a_j^+] = 0$$

and for fermions:

$$[a_i, a_j^+]_+ = \delta_{ij},\quad [a_i, a_j]_+ = [a_i^+, a_j^+]_+ = 0$$

However, they share the same commutation:

$$[a_i^+ a_j, a_k^+ a_l] = \delta_{jk} a_i^+ a_l - \delta_{il} a_k^+ a_j$$

For k fermions in r- orbit space, the reference state (the ground state is chosen) is $|\overbrace{1,\ldots 1}^{k},\overbrace{0,\ldots 0}^{r-k}\rangle$. For the r-orbit space, we can construct the operator set $\{a_i^+ a_j,\ 1 \le i, j \le r\}$. These operators satisfy the above relation and they constitute u(r) (a unitary algebra). Meanwhile, $\{a_i^+ a_j\}_{1\le i,j\le k}$ and $\{a_i^+ a_j\}_{k+1\le i,j\le r}$ can annihilate the reference state and they form the subalgebras u(k) and u(r-k), respectively. Hence, the coset space is U(r)/(U(k) × U(r-k)). For the system of a single electron, we have

U(r)/(U(1) × U(r-1)). This space is equivalent to SU(r)/U(r-1) which is an r dimensional hypersphere S^r.

For the bosonic system, the reference state is $\mid n,\overbrace{0,...0}^{r-1}\rangle$, the corresponding Lie algebra and subalgebras are u(r), u(1) and u(r-1), respectively. The coset space is U(r)/(U(1)×U(r-1)) which is SU(r)/ U(r-1) ~ S^r.

Therefore, the main difference of fermions and bosons lies in their different reference states which originate from fermions having to obey Pauli's exclusion principle while for bosons, multiple ones are allowed in the same state. Besides this difference, interestingly, the algebraic treatment of one electron system is just like that for the bosons.

2.5 Explicit expressions

In this section, the explicit expressions for some of the Lie operators will be demonstrated for later usage.

A. su(2)

For the system of two oscillators, we have $J_+ = a_1^+ a_2$, $J_- = a_2^+ a_1$, $J_z = \frac{1}{2}(a_1^+ a_1 - a_2^+ a_2)$ and {J_z, J_+, J_-} form an su(2) algebra: $[J_z, J_\pm] = \pm J_\pm$, $[J_+, J_-] = 2J_z$. {J_z} is u(1). The coset space is SU(2)/U(1). The coset representatives are

$$\begin{aligned}|J,\Omega\rangle &= \exp(\tau J_+ - \tau^* J_-)|J,-J\rangle \\ &= N_{or}\exp(\tau J_+)|J,-J\rangle \\ &= N_{or}\sum_n \frac{(\tau J_+)^n}{n!}|J,-J\rangle \\ &= N_{or}\sum_n \frac{\tau^n}{n!} n! \begin{bmatrix} 2J \\ n \end{bmatrix}^{1/2} |J,n-J\rangle\end{aligned}$$

Here, $\begin{bmatrix} 2J \\ n \end{bmatrix} \equiv \frac{(2J)!}{n!(2J-n)!}$. Since $[J_+, J_-] \neq 0$, therefore,

$$\exp\ (\tau J_+ - \tau^* J_-) = \mathrm{N_{or}} \exp(\tau J_+)\exp(-\tau^* J_-)$$

and

$$\exp(-\tau^* J_-)|J,-J\rangle = |J,-J\rangle$$

By the normalization condition of $|J,\Omega\rangle$, $\langle J,\Omega|J,\Omega\rangle = 1$. N_{or} can be determined.

Since

$$\langle J,\Omega| = N_{or} \sum_{n'} \frac{(\tau^*)^{n'}}{n'!} n'! \begin{bmatrix} 2J \\ n' \end{bmatrix}^{1/2} \langle J, n' - J|$$

therefore,

$$\langle J,\Omega|J,\Omega\rangle = N_{or}^2 \sum_n (|\tau|^2)^n \begin{bmatrix} 2J \\ n \end{bmatrix} = N_{or}^2\ (1+|\tau|^2)^{2J}$$

and $N_{or} = \frac{1}{(1+|\tau|^2)^J}$. In the above process, we have used two expressions which will be encountered repeatedly afterwards:

$$e^x = \sum \frac{x^n}{n!}$$

and

$$(x+y)^p = \sum \begin{bmatrix} p \\ n \end{bmatrix} x^{p-n} y^n$$

Analogously, we have:

$$\langle J,\Omega|n_1|J,\Omega\rangle = \frac{2J|\tau|^2}{1+|\tau|^2} \quad (n_1 = a_1^+ a_1)$$

$$\langle J,\Omega|n_2|J,\Omega\rangle = \frac{2J}{1+|\tau|^2} \quad (n_2 = a_2^+ a_2)$$

$$\langle J,\Omega|n_1^2|J,\Omega\rangle = \frac{2J|\tau|^2 (1+2J|\tau|^2)}{(1+|\tau|^2)^2}$$

$$\langle J,\Omega|n_2^2|J,\Omega\rangle = \frac{2J(2J+|\tau|^2)}{(1+|\tau|^2)^2}$$

$$\langle J,\Omega|n_1 n_2|J,\Omega\rangle = \frac{2J(2J-1)|\tau|^2}{(1+|\tau|^2)^2}$$

$$\langle J,\Omega|J_+|J,\Omega\rangle = \frac{2J\tau^*}{1+|\tau|^2}$$

$$\langle J,\Omega|J_-|J,\Omega\rangle = \frac{2J\tau}{1+|\tau|^2}$$

$$\langle J,\Omega|J_z|J,\Omega\rangle = \frac{J(|\tau|^2-1)}{1+|\tau|^2}$$

$$\langle J,\Omega|J_+{}^2|J,\Omega\rangle = \frac{2J(2J-1)(\tau^*)^2}{(1+|\tau|^2)^2}$$

$$\langle J,\Omega|J_-{}^2|J,\Omega\rangle = \frac{2J(2J-1)\tau^2}{(1+|\tau|^2)^2}$$

Here, we should note that, say, $\langle J,\Omega|n_1{}^2|J,\Omega\rangle$ is not equal to $\langle J,\Omega|n_1|J,\Omega\rangle^2$ and also that $\langle J,\Omega|n_1 n_2|J,\Omega\rangle$ is not equal to $\langle J,\Omega|n_1|J,\Omega\rangle\langle J,\Omega|n_2|J,\Omega\rangle$. All these demonstrate that the coset representation does show up the quantal characteristic!

Since the coset space is S^2, τ can be written in terms of the generalized coordinates (q, p) as:

$$\tau = (q+ip)/[J-(q^2+p^2)]$$

then, we have the following expressions which are useful in the study of two-mode system.

$$n_1(\Omega) = 2(q^2+p^2), \quad n_2(\Omega) = 2[J-(q^2+p^2]\,,$$

$$n_1{}^2(\Omega) = 2(q^2+p^2)[\frac{J-(q^2+p^2)}{J} + 2(q^2+p^2)]$$

$$n_2{}^2(\Omega) = 2J[2J + \frac{(q^2+p^2)}{J-(q^2+p^2)}][\frac{J-(q^2+p^2)}{J}]^2$$

$$n_1 n_2(\Omega) = \frac{2(2J-1)}{J}[J-(q^2+p^2)](q^2+p^2)$$

$$(J_+ + J_-)(\Omega) = 4q[J-(q^2+p^2)]^{1/2}$$

$$J_z = 2(q^2 + p^2) - J$$

$$(J_+^{\ 2} + J_-^{\ 2})(\Omega) = 4(2 - \frac{1}{J})(q^2 + p^2)[J - (q^2 + p^2)]$$

In these expressions, note that $(q^2 + p^2) < J$. Besides, (q, p) can be related to the polar coordinates (θ, ϕ) of S^2 :

$$q = \sqrt{J} \sin\frac{\theta}{2} \cos\phi \,, \quad p = -\sqrt{J} \sin\frac{\theta}{2} \sin\phi$$

J_x^2, J_y^2, J_z^2 are related to molecular rotation. Their expressions are:

$$J_z^{\ 2}(\Omega) = [2(q^2 + p^2) - J]^2 + \underline{2J^{-1}[J - (q^2 + p^2)](q^2 + p^2)}$$

$$J_x^2(\Omega) = u + v \,, \quad J_y^2(\Omega) = -u + v$$

$$u = (2 - \underline{J^{-1}})(q^2 - p^2)[J - (q^2 + p^2)]$$

$$v = \frac{1}{2}\{4[J - (q^2 + p^2)](q^2 + p^2) + \underline{[J - (q^2 + p^2)]^2 J^{-1} + (q^2 + p^2)^2 J^{-1}}\}$$

The underlined parts are by the quantal effect. ($J_x^2 + J_y^2 + J_z^2$) (Ω) can be written as two parts: one is the classical part and the other is due to the quantal effect. They are J^2 and J, respectively. Therefore, the J term in $J(J+1)$, which is the eigenvalue of $\hat{J}^2$, is indeed from the quantal effect. Obviously, coset representatives can show up the quantal effect of rotation. This result can hardly be elucidated by other algorithms.

B. su(3)

For the system of three oscillators, we have $I_+ = a_1^+ a_2$, $I_- = a_2^+ a_1$, $J_+ = a_1^+ a_3$, $J_- = a_3^+ a_1$, $K_+ = a_2^+ a_3$, $K_- = a_3^+ a_2$, $I_z = (a_1^+ a_1 - a_2^+ a_2)/2$, $Y = (a_1^+ a_1 + a_2^+ a_2 - 2a_3^+ a_3)/2$. They form an su(3) algebra:

$$[I_+, I_-] = 2I_z \,, \quad [I_z, I_\pm] = \pm I_\pm \,, \quad [I_z, J_\pm] = \pm J_\pm /2$$

$$[I_z, K_\pm] = \mp K_\pm /2 \,, \quad [J_+, J_-] = I_z + Y \,, \quad [K_+, K_-] = Y - I_z$$

$$[J_\pm, K_\pm] = 0,\ [Y, J_\pm] = \pm 3/2 J_\pm,\ [Y, K_\pm] = \pm 3/2 K_\pm$$

su(3) contains an su(2) subalgebra. The reference state is $|N,0,0\rangle$ which is the state with $n_1 = N$, $n_2 = n_3 = 0$.

Note that $|N,0,0\rangle$ is the eigenstate of I_z, Y, K_+, K_- ($K_\pm|N,0,0\rangle = 0$) which form a u(2) algebra. The coset space is SU(3)/U(2). The coset representatives are:

$$|\Lambda\Omega\rangle = \exp(\tau_1 I_+ + \tau_2 J_+ - \tau_1^* I_- - \tau_2^* J_-)|N,0,0\rangle$$

$$= N_{or} \exp(-\tau_1^* I_- - \tau_2^* J_-)|N,0,0\rangle$$

As before in the case of su(2), N_{or} can be calculated as

$$(1+\tau^+\tau)^{-N/2} (\tau^+\tau = |\tau_1|^2 + |\tau_2|^2)$$

Also that:

$$\langle\Lambda\Omega|n_1|\Lambda\Omega\rangle = N\frac{1}{1+\tau^+\tau}$$

$$\langle\Lambda\Omega|n_2|\Lambda\Omega\rangle = N\frac{|\tau_1|^2}{1+\tau^+\tau}$$

$$\langle\Lambda\Omega|n_3|\Lambda\Omega\rangle = N\frac{|\tau_2|^2}{1+\tau^+\tau}$$

$$\langle\Lambda\Omega|I_+|\Lambda\Omega\rangle = N\frac{(-\tau_1^*)}{1+\tau^+\tau}$$

$$\langle\Lambda\Omega|J_+|\Lambda\Omega\rangle = N\frac{(-\tau_2^*)}{1+\tau^+\tau}$$

$$\langle\Lambda\Omega|K_+|\Lambda\Omega\rangle = N\frac{\tau_1\tau_2^*}{1+\tau^+\tau}$$

$$\langle\Lambda\Omega|I_z|\Lambda\Omega\rangle = \frac{N}{2}\frac{1-2|\tau_1^2|-|\tau_2|^2}{1+\tau^+\tau}$$

$$\langle\Lambda\Omega|Y|\Lambda\Omega\rangle = \frac{N}{2}\frac{1-3|\tau_2|^2}{1+\tau^+\tau}$$

(Those for I_-, J_-, K_- are the complex conjugates of the above expressions respectively.)

Note that though $\langle\Lambda\Omega|n_1 n_2|\Lambda\Omega\rangle$ can be written as the product of $\langle\Lambda\Omega|n_1|\Lambda\Omega\rangle$ and $\langle\Lambda\Omega|n_2|\Lambda\Omega\rangle$, the coefficient is $N(N-1)$, not N^2. This is also true for other cases like $\langle\Lambda\Omega|J_+K_+|\Lambda\Omega\rangle$.

Let us define :

$$I_x=(I_++I_-)/2,\quad I_y=-\frac{i}{2}(I_+-I_-),\quad I_z=(n_1-n_2)/2$$

$$J_x=(J_++J_-)/2,\quad J_y=-\frac{i}{2}(J_+-J_-),\quad J_z=(n_1-n_3)/2$$

$$K_x=(K_++K_-)/2,\quad K_y=-\frac{i}{2}(K_+-K_-),\quad K_z=(n_2-n_3)/2$$

The generalized coordinates (q_1, p_1, q_2, p_2) are chosen as

$$-\tau_1^*=(q_1+ip_1)/[2N-\sum (q_i^2+p_i^2)]^{1/2}$$

$$-\tau_2^*=(q_2+ip_2)/[2N-\sum (q_i^2+p_i^2)]^{1/2}$$

Then we have

$$I_x=\frac{1}{2}q_1[2N-\sum (q_i^2+p_i^2)]^{1/2},\quad I_y=\frac{1}{2}p_1[2N-\sum (q_i^2+p_i^2)]^{1/2}$$

$$I_z=\frac{1}{2}[N-(q_1^2+p_1^2)-\frac{1}{2}(q_2^2+p_2^2)]$$

$$J_x=\frac{1}{2}q_2[2N-\sum (q_i^2+p_i^2)]^{1/2},\quad J_y=\frac{1}{2}p_2[2N-\sum (q_i^2+p_i^2)]^{1/2}$$

$$J_z=\frac{1}{2}[N-\frac{1}{2}(q_1^2+p_1^2)-(q_2^2+p_2^2)]$$

$$K_x=\frac{1}{2}(q_1q_2+p_1p_2),\quad K_y=\frac{1}{2}(q_1p_2-q_2p_1)$$

$$K_z=\frac{1}{4}(q_1^2+p_1^2-q_2^2-p_2^2)$$

$$n_1=N-\frac{1}{2}\sum (q_i^2+p_i^2),\quad n_2=\frac{1}{2}(q_1^2+p_1^2),\quad n_3=\frac{1}{2}(q_2^2+p_2^2)$$

These expressions are very useful for the study of a three-mode vibrational system.

C. su(3) (single electron system)

It was mentioned previously that the coset space for a single electron in a three-orbit space is SU(3)/U(2). The reference state is $|1,0,0\rangle$ and

$$|\Omega\rangle = \exp(\tau_{21}a_2^+a_1 + \tau_{31}a_3^+a_1 - h.c.)|1,0,0\rangle$$
$$= N_{or}(1 + \sum_{j=2,3}\tau_{j1}a_j^+a_1)|1,0,0\rangle$$

N_{or} can be obtained as $(1+|\tau_{21}|^2+|\tau_{31}|^2)^{-1/2}$. For convenience, we choose $\tau_1 = \tau_{21}$, $\tau_2 = \tau_{31}$ and the generalized coordinates q_1, p_1, q_2, p_2 as

$$\tau_1 = (q_1 + ip_1)/[1 - \sum (q_i^2 + p_i^2)]^{1/2}$$
$$\tau_2 = (q_2 + ip_2)/[1 - \sum (q_i^2 + p_i^2)]^{1/2}$$

then

$$N_{or} = [1 - \sum (q_i^2 + p_i^2)]^{1/2}$$

and by calculation, we have

$$(\sum_i e_i a_i^+ a_i)(\Omega) = N_{or}^2(e_1 + e_2|\tau_1|^2 + e_3|\tau_2|^2)$$
$$= e_1[1 - \sum (q_i^2 + p_i^2)] + e_2(q_1^2 + p_1^2) + e_3(q_2^2 + p_2^2)$$

and

$$(\sum \frac{1}{2} V_{kl} a_k^+ a_l)(\Omega) = N_{or}^2(V_{12} + V_{13}\tau_2 + V_{23}\tau_1^*\tau_2 + c.c)$$
$$= (V_{12}q_1 + V_{13}q_2)\,[1 - \sum (q_i^2 + p_i^2)]^{1/2} + V_{23}(q_1q_2 + p_1p_2)$$

D. so(6)

The electronic correlation can be of interest. The simplest system is two electrons in three-orbit space. Its Lie algebra is so(6) with operators: $\{a_i^+a_j - \frac{1}{2}\delta_{ij}(1 \le i,j \le 3), a_i a_j, a_i^+ a_j^+ (1 \le i \ne j \le 3)\}$. Its subalgebra is u(3) with $\{a_i^+ a_j (1 \le i,j \le 3)\}$. The coset space is SO(6)/U(3). The reference state is $|0,0,0\rangle$ and the coset representatives are

$$|\Omega\rangle = \exp(\tau_{ij}a_i^+a_j - h.c.)|0,0,0\rangle$$
$$= N_{or}\exp(\sum \tau_{ij}a_i^+a_j^+)|0,0,0\rangle$$
$$= N_{or}(1 + \sum \tau_{ij}a_i^+a_j^+)|0,0,0\rangle$$

N_{or} is $(1+2\sum|\tau_{ij}|^2)^{-1/2}$.

With $\tau_1 = \tau_{23}$, $\tau_2 = \tau_{13}$, $\tau_3 = \tau_{12}$

and

$$\tau_i = \frac{1}{\sqrt{2}}(q_i + ip_i)/[1 - \frac{1}{2}\sum(q_i^2 + p_i^2)]^{1/2}$$

then,

$$N_{or} = [\frac{1 - \frac{1}{2}\sum(q_i^2 + p_i^2)}{1 + \frac{3}{2}\sum(q_i^2 + p_i^2)}]^{1/2}$$

$$(\sum_i e_i a_i^+ a_i)(\Omega)$$

$$= (1+2\sum|\tau_i|^2)^{-1} 4\{|\tau_1|^2 (e_2 + e_3) + |\tau_2|^2 (e_1 + e_3) + |\tau_3|^2 (e_1 + e_2)\}$$

$$= 2[1 + \frac{3}{2}\sum(q_i^2 + p_i^2)]^{-1}[(q_1^2 + p_1^2)(e_2 + e_3) + (q_2^2 + p_2^2)(e_1 + e_3) + (q_3^2 + p_3^2)(e_1 + e_2)]$$

and

$$(\sum \frac{1}{2} V_{ij} a_i^+ a_j)(\Omega)$$

$$= [1+2\sum|\tau_i|^2]^{-1} 4\,\mathrm{Re}(V_{12}\tau_1\tau_2^* + V_{23}\tau_2\tau_3^* - V_{13}\tau_1\tau_3^*)$$

$$= 2[1 + \frac{3}{2}\sum(q_i^2 + p_i^2)]^{-1}[V_{12}(q_1q_2 + p_1p_2) + V_{23}(q_2q_3 + p_2p_3) - V_{13}(q_1q_3 + p_1p_3)]$$

Let $e_i = 1$, then $\sum a_i^+ a_i$ shows the number of electrons that are paired. Its value is

$$4\sum(q_i^2 + p_i^2)[1 + \frac{3}{2}\sum(q_i^2 + p_i^2)]^{-1}$$

Since $0 \le \sum(q_i^2 + p_i^2) \le 2$, the number of pairing electrons is between 0 and 2. This is exactly as expected.

E. su(1, 1)

The coset spaces that are related to the molecular stationary states are compact with finite space such as SU(2)/U(1), SU(3)/U(2). The noncompact coset spaces correspond to systems with unbounded space. They are sometimes useful. SU(1,1)/U(1) is such a case.

su(1,1) consists of $\{K^0, K^+, K^-\}$ with the following properties:

$$[K^0, K^\pm] = \pm K^\pm, \quad [K^+, K^-] = -2K^0, \quad K^2 = K^{02} - \frac{1}{2}(K^+K^- + K^-K^+)$$

$$K^+|k,n\rangle = [(n+1)(2k+n)]^{1/2}|k,n+1\rangle$$

$$K^-|k,n\rangle = [n(2k+n-1)]^{1/2}|k,n-1\rangle$$

$$K^0|k,n\rangle = (n+k)|k,n\rangle$$

$$K^2|k,n\rangle = k(k-1)|k,n\rangle$$

The eigenstates are $|k,n\rangle$ with $n = 0, 1, 2,\ldots$ $k = 1/2, 1, 3/2,\ldots$

$\{K^0\}$ forms the subalgebra u(1). The coset space is SU(1,1)/U(1) which is a hyperbolic surface. The coset representatives are:

$$|k,\Omega\rangle = \exp(\tau K^+ - \tau^* K^-)|k,0\rangle$$

$$= N_{or}\exp(\tau K^+)|k,0\rangle$$

N_{or} can be calculated as $(1-|\tau|^2)^k$

$$\langle k,\Omega|K^+|k,\Omega\rangle = \frac{2k\tau^*}{1-|\tau|^2}$$

$$\langle k,\Omega|K^-|k,\Omega\rangle = \frac{2k\tau}{1-|\tau|^2}$$

$$\langle k,\Omega|K^0|k,\Omega\rangle = k\frac{1+|\tau|^2}{1-|\tau|^2}$$

Taking $\tau = \dfrac{q+ip}{[4k+q^2+p^2]^{1/2}}$,

then

$$K^+(k\Omega) = \frac{1}{2}(q-ip)[4k+q^2+p^2]^{1/2}$$

$$K^-(k\Omega) = \frac{1}{2}(q+ip)[4k+q^2+p^2]^{1/2}$$

$$K^0(k\Omega) = k + \frac{1}{2}(q^2+p^2)$$

In application, k is often chosen as 1/2.

F. h(4)

h(4) is composed of $\{a^+, a, a^+a, I\}$. It is suitable for describing a photon system. Its subalgebras are $\{a^+a\}$ and $\{I\}$. Both are u(1). The coset space is H(4)/U(1)× U(1) which is an one dimensional complex plane. The coset representatives are:

$$|\Omega\rangle = \exp(za^+ - z^* a)|0\rangle$$

$$= N_{or} \exp(za^+)|0\rangle$$

It can be calculated that

$$N_{or} = e^{-|z|^2/2}$$

and

$$a(\Omega) = z$$

$$a^2(\Omega) = z^2$$

z can be written as $q + ip$. Here, we should note that $|a(\Omega)|^2 = a^2(\Omega)$

2.6 Heisenberg's correspondence

Heisenberg's concept of a quantal system is that its dynamical quantities can be described by matrices of which the diagonal elements are the quantities of its stationary states while the off-diagonal elements are the transitional or interacting quantities among its states. In the realm of large quantum numbers, these off-diagonal terms can correspond to a set of action-angle variables. For example, for two interacting states *s* and *t*, their interaction can be described by the creation and destruction operators. Each operator can be correspondent to action n and angle ϕ. In other words, classically, we have :

$$a_s^+ \sim \sqrt{n_s} e^{i\phi_s}, \quad a_t^+ \sim \sqrt{n_t} e^{i\phi_t}$$

Note that these action/angle analogues will be adopted regardless of harmonic and Morse oscillators. The generalized coordinates can be chosen as:

$$q_\alpha = \sqrt{2n_\alpha}\cos\phi_\alpha, \quad p_\alpha = -\sqrt{2n_\alpha}\sin\phi_\alpha \quad (\alpha = s,t)$$

and

$$a_s^+ a_t + a_t^+ a_s \sim q_s q_t + p_s p_t$$

(Note other choices are also possible like $q_\alpha = \sqrt{n_\alpha/2}\cos\phi_\alpha$, $p_\alpha = -\sqrt{n_\alpha/2}\sin\phi_\alpha$)

In many cases, such as su(2) and su(3), the expressions obtained in this way are the same as those under the classical limit in the coset representation. (For a noncompact system such as so(6), there will be difficulty in having this equivalency). Hence, if we are not so much concerned with the quantal effect, Heisenberg's correspondence can be adopted for convenience. Herein, we see Heisenberg's deep insight into quantum mechanics.

For the above mentioned interaction, since the total action $n_s + n_t$ is conserved, we can also adopt

$$a_s^+ \sim \sqrt{n_s}, \quad a_t^+ \sim \sqrt{n_t}\, e^{i\phi_t}$$

Here, ϕ_t is the phase difference between *s* and *t* actions. Thereby, we have

$$a_s^+ a_t + a_t^+ a_s \sim \sqrt{2n_s}\, q_t$$

In other words, as there is conservative quantity, we can adopt this algorithm for convenience. In fact, in the previous expressions with the coset representatives like SU(3)/U(2) and SU(2)/U(1), there are conserved quantities *N* and *J*. This is why their spaces are finite (S^3 and S^2 spheres) and compact.

References

2. 1 W. M. Zhang, D. H. Feng and R. Gilmore, *Rev. Mod. Phys.*, 62(1990)867.

2. 2 R. Gilmore, Lie groups, *Lie Algebras and Some of their Applications*, Wiley, New York, 197

Chapter 3

Concepts of nonlinear dynamics

3.1 Universality of chaos

Our world is nonlinear. As an approximation, we sometimes use a linear method to describe nonlinear properties. However, we have to know the limit of this linear approximation. Nonlinearity is fundamentally different from linearity. The core characteristics of nonlinearity cannot be realized through linear approximation.

Nonlinearity is the origin of chaos. We know that chaos does not need a complicated cause. A very simple cause may lead to very complicated chaos. (Here, we do not intend to define chaos in a strict way.).

Consider an N dimensional space in which the equations of motion are:

$$\dot{x}_1 = F_1(x_1,...,x_N)$$

$$\dot{x}_2 = F_2(x_1,...,x_N)$$

$$\cdots$$

$$\dot{x}_N = F_N(x_1,...,x_N)$$

Here, $(x_1,...,x_N)$ are the coordinates of a point at time t. F_i are the functional forms the coordinate components have to follow. We denote the position of a point at t as $\vec{x}(t)$. For a given initial point $\vec{x}(0)$, then the position at any later time $\vec{x}(t)$ can be known via the integration of F_i. Similar to the set of differential equations, we have the map:

$$\vec{X}_{n+1} = M(\vec{X}_n)$$

M is the map transformation. Index n shows the iterations of the map. The trajectory can be denoted as:

$$\vec{X}_0,\ \vec{X}_1,\ \ldots,\ \vec{X}_n, \ldots$$

This concept of trajectory is very useful. In fact, some differential equations may be cast into the form as shown above. For instance, for a forced damped pendulum, the differential equation of motion is:

$$\ddot{\theta} + v\dot{\theta} + \sin\theta = T\sin\omega t$$

Let $x_1=\dot{\theta}$, $x_2= \theta$, $x_3=\omega t$

then

$$\dot{x}_1 = T\sin x_3 - \sin x_2 - v x_1$$

$$\dot{x}_2 = x_1$$

$$\dot{x}_3 = \omega$$

This equation set is exactly what we had previously with $N=3$.

For this kind of equation set, theory tells us that as $N \geq 3$, trajectories can be chaotic while for $N = 1$, 2, there will be no chaotic trajectories. For the forced damped pendulum, $N=3$. Hence, its motion can be chaotic. For a map, if it is invertible and with $N \geq 2$, then its trajectories can be chaotic. Otherwise, if $N = 1$, the motion is regular and ordered. If the map is noninvertible, then even with $N = 1$, the motion can be chaotic. In summary, we know that chaotic motion is universal.

Chaotic motion possesses two basic characteristics: one is that augmentation and compression coexist. The other is that its trajectories are very initial point dependent. Suppose initially, two points have a deviation $\Delta(0)$ and at t the deviation is $\Delta(t)$. If

$$\Delta(t) \sim \Delta(0)e^{ht}$$

and h>0, then the trajectories are very initial point dependent. Now, it is generally accepted as the definition that a trajectory is chaotic if it follows a deterministic equation of motion and its h, the Lyapunov exponent, is larger than 0.

For example, for a binary shift map, error is augmented 2^n fold after n maps. Since

$$2^n = e^{n\ln 2}$$

then, $h = \ln 2$. If the initial accuracy is 10^{-14}, then after $n=$ 45 iterations, the accuracy is only 1. This is because that we have:

$$2^{45} \cdot 10^{-14} \sim 1$$

3.2 One-dimensional map

1-dimensional tent map is shown in Fig. 3.1:

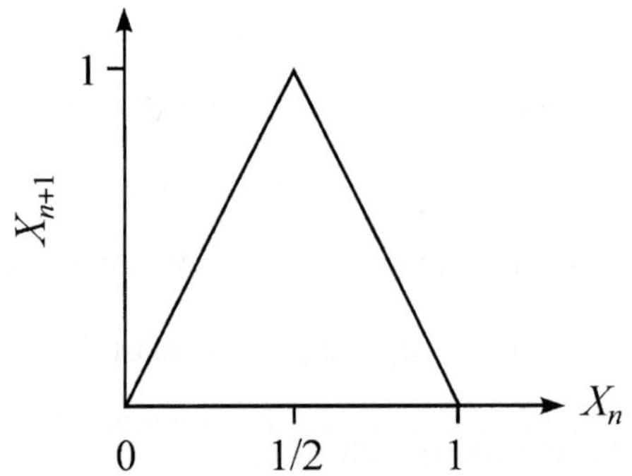

Fig.3.1 The tent map

This transformation is to map the points in [0,1] according to the rules:

As $X_n < 1/2$, $X_{n+1} = 2X_n$

As $X_n > 1/2$, $X_{n+1} = 2(1 - X_n)$

This transformation augments 2-fold any number less than 1/2 while it compresses a number greater than 1/2 (after first augmentation).

After *m* mappings, the relation between X_n and X_{n+m} is as shown in Fig.3.2:

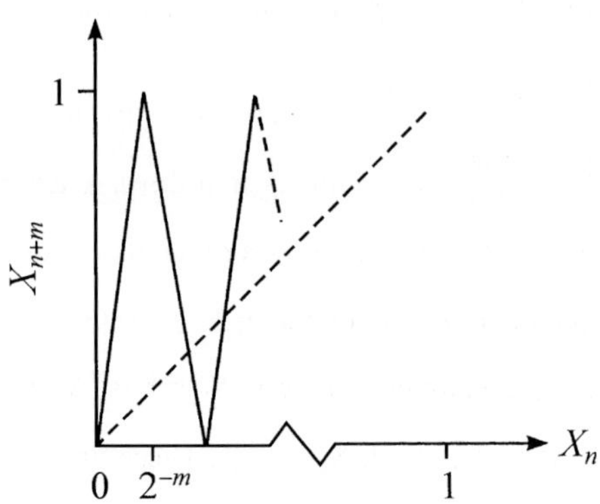

Fig.3.2 The map from X_n to X_{n+m}. The dash line is the diagonal.

The significance of Fig.3.2 is that two points within a deviation of 2^{-m} will be scattered anywhere in [0, 1] after m mappings. That is, a point in [0, 1], after many iterations of mapping, will be scattered around *chaotically* in [0, 1]. The Lyapunov exponent is ln2.

In [0, 1], besides chaotic points, there are numerous periodic points. It is easy to find them. As shown in Fig.3.2, those points that intersect with the diagonal line are unmoved after m mappings and are of period m. Their number is 2^m. Of course, among them, there are pseudo period-m points such as $X_n = 0$ which is in fact a fixed point under the map. Apparently, the point of any large periods can exist. The number of periodic points is infinite and they can be in one-to-one correspondence with the positive numbers, i.e., they are countable. The number of chaotic points is infinite. However, they are uncountable.

Periodic points are dense in [0, 1] and can be anywhere. In Fig.3.2, for a very small interval $[\varepsilon, 2\varepsilon]$, as long as p is large enough so that $2^{-p} < \varepsilon/2$, then in $[\varepsilon, 2\varepsilon]$, there will be a point of period p. In summary, periodic points and chaotic points are mingled together, just as rational and irrational numbers are in [0, 1].

The probability that a randomly chosen point in [0, 1] is periodic is 0. It has to be a chaotic point. Then, after many mappings, its trace will be evenly distributed in [0, 1]. Therefore, we define the distribution as the natural invariant density for the map which is $\rho(x) = 1$ in this case. For other cases, ρ may not be a constant. ρ is an intrinsic property of a map. It is an invariance of the map.

Another famous map is the logistic map:

$$X_{n+1} = rX_n(1 - X_n), \quad X_n \text{ in } [0, 1]$$

Note that the chaotic property of this map depends on r.

As r<3.57 ..., the points in [0, 1] are attracted to periodic orbits. As r increases, the periodic orbits bifurcate from period 1 to 2, 4, 8, ..., etc. These orbits are stable with negative Lyapunov exponents. Besides, there are unstable periodic orbits.

As $r>3.57...$, the system runs into a chaotic state. This chaotic state possesses a very complicated structure in which there are bifurcations of periodic orbits and period-3 orbits like windows and many more. A unique characteristic of this structure is self-similarity. Its parts are similar to the whole and the parts in the parts are similar to the parental parts in an infinite way.

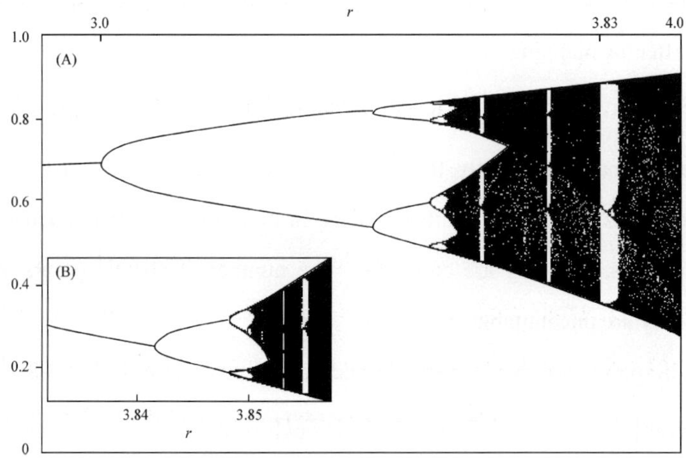

Fig.3.3 The logistic map The inset (B) is a magnification of period-3 window

Fig.3.3 shows the logistic behavior for various values of r. The procedure is to take a random number in [0, 1] and record its value after 500 iterations. In the Figure, we see the bifurcation of periodic orbits. The inset is the enlargement of the period-3 window in which there are the bifurcations of periodic orbits, chaotic orbits and windows, showing self-similarity. We note that unstable periodic orbits are hardly shown up in the Figure, numerically.

The logistic map possesses many more complicated properties. We do not intend to cover all these since we are concerned more with the general concept of nonlinearity in this chapter.

The properties of the 1-dimensional maps shown above are universal. They are the common characteristics of nonlinearity and are very useful for our understanding of nonlinear phenomena in physics.

3.3 Period 3 implies chaos

The direct way of judging a chaotic system is to calculate whether its Lyapunov exponents are positive. However, the Sarkovskii theorem tells us that the appearance of periodic trajectories can also judge this.

Sarkovskii arranged the positive integers as follows. First the odd numbers except 1, then 2 times, 2^2 times, ..., these numbers. Finally, the powers of 2 in a decreasing order and 1, i.e.,

$$3,\ 5,\ 7,\ \ldots,\ 2\times3,\ 2\times5,\ 2\times7,\ \ldots,\ 2^2\times3,\ 2^2\times5,\ 2^2\times7,\ \ldots,\ 2^5,\ 2^4,\ 2^3,\ 2^2,\ 2,\ 1$$

The sequence is called Sarkovskii sequence.

The Sarkovskii theorem tells us that in a 1-dimensional system under a continuous map, if a trajectory of period p appears then the trajectories of periods that are after p in the Sarkovskii sequence will also appear. Hence, if an odd number period appears, then any periods will appear except those ahead of the odd number in the sequence which will not necessarily appear. In other words, we will have period-infinity, or chaotic trajectories. (Conceptually, we may say that a chaotic trajectory is of period-infinity or just of no period.)

Obviously, by Sarkovskii sequence, the appearance of period-3 trajectory implies the appearance of trajectories of all periods. That is, 'period 3 implies chaos'.

3.4 KAM theorem

For an energy conserved Hamiltonian system, the volume of a subspace in its phase space is conserved as time elapses. However, its shape may vary. Therefore, there can be augmentation and compression for the subspace, or its dynamics can be chaotic.

Suppose the phase space of a Hamiltonian is of dimension $2N$ and there are N conserved quantities, then the system is integrable since its motion can be decomposed into N independent degrees of freedom. The dynamics of an integrable system can be described by the appropriately chosen conserved actions I_i and their angles θ_i (i=1, ... N). The system Hamiltonian H_0 can be described as the function of I_i: $H_0(I_1,...I_N)$. The equations of motion, or the Hamilton's equations of motion, are:

$$\partial H_0(I_1,...I_N)/\partial I_i = \dot{\theta}_i \equiv \omega_{0i}(I_i)$$

$$\partial H_0(I_1,...I_N)/\partial \theta_i = -\dot{I}_i = 0$$

The second equation shows that I_i is conserved. Since ω_{0i} is independent of time, from the first equation, we have:

$$\theta_i(t) = \theta_i(0) + \omega_{0i}(I_i)t$$

The motion is the very simple harmonic motion.

In general, ω_{0i} is related to I_i, i.e., action will effect frequency. This is the meaning of nonlinearity.

If $\{\omega_{01},...,\omega_{0N}\}$ are in rational ratios, then the motion is still periodic. Otherwise, it is quasi-periodic.

The dynamics of an integrable system is simple. While under perturbations, εH_1, the conserved quantities will be destroyed. What are the system dynamics of $H = H_0 + \varepsilon H_1$? This problem is very complicated. Kolmogorov, Arnold and Moser answered the question in the following picture. This is the KAM theorem.

As εH_1 increases, the trajectories of which $\{\omega_{01},...,\omega_{0N}\}$ are in rational ratios will be destroyed first and bifurcate into chaotic trajectories, stable and unstable periodic trajectories with new frequencies that are in rational or irrational ratios. Those trajectories with irrational ratios of $\{\omega_{01},...,\omega_{0N}\}$ will persist until the perturbation εH_1 is larger. The newly born periodic trajectories will be destroyed in a similar way with the birth of more chaotic and periodic trajectories. This is an infinite self-similarity process.

Fig.3.4 shows this process diagrammatically in which (a) is the integrable structure. The center is the fixed point. Solid circles show the regions of which $\{\omega_{01},...,\omega_{0N}\}$ are of irrational ratios while the dashed circles show the regions of which $\{\omega_{01},...,\omega_{0N}\}$ are of rational ratios. (b) is the structure by KAM theorem in which solid circles remain though they can be distorted somehow, while the dashed circles are destroyed with the birth of new stable (the concentric ones form an orbit), unstable periodic and chaotic trajectories.

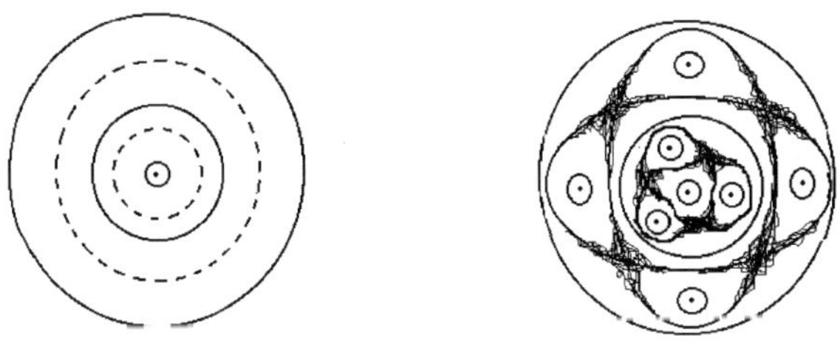

(a) (b)

Fig.3.4 The KAM theorem. (a) The integrable structure. (b) The destroyed structure by KAM theorem. Points show the stable trajectories. Unstable trajectories are in the chaotic region. See text for details.

In summary, as the perturbation εH_1 increases, more periodic and quasi-periodic trajectories are destroyed with more chaotic trajectories showing up.

In Chapter 9, we will address the KAM theorem again.

3.5 Poincare surface of section

The observation of a multidimensional trajectory $(q_1, q_2, ..., q_N, p_1, p_2, ...p_N)$ is not convenient. Poincare devised a surface of section to observe it. By the conservation of energy, a trajectory will move in a $2N-1$ dimensional space. Then, if we slice a surface of section, say, along $q_1 = K_0$ to observe it, the trajectory will simply be:

$$\vec{X} = (q_2, ..., q_N, p_2, ...p_N)$$

Here, $(q_2, ..., q_N, p_2, ...p_N)$ are the coordinates of $\vec{X}$ on the surface of the section. That is, every time the trajectory pierces the surface, a trace will remain on it. On the surface, we will see a series of points, $\vec{X}_n$ (n = 0,1,2,...). Will $\vec{X}_{n+1}$ uniquely defined by $\vec{X}_n$ on the surface? The answer is no! This is because that Hamiltonian or energy is a function of p_1^2. The conservation of energy simply cannot ensure the uniqueness of p_1. For $\vec{X}_{n+1}$ to be uniquely determined by $\vec{X}_n$, we will require that only when $p_1>0$ (or <0) as the trajectory pierces the surface, will it leave a trace. By this compensation, we will observe a series of uniquely defined points on the surface. These points can reliably reflect the property of the trajectory.

By Poincare surface of section, the dynamical space is reduced by one degree of freedom, however, this will bring us much convenience.

3.6 Kicked rotor

The Hamiltonian of a rotor with rotational inertia, I, under periodic impulses of period τ as shown in Fig.3.5 is:

$$H(p_\theta,\theta,t) = \frac{p_\theta^2}{2I} + k\cos\theta\sum_m \delta(t-m\tau)$$

Here, p_θ is angular momentum. δ is the Dirac function.

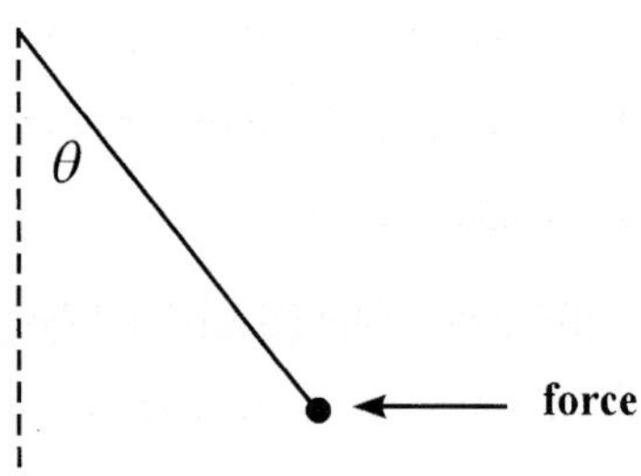

Fig.3.5 The rotor under periodic impulses

The equations of motion are:

$$\dot{p}_\theta = k\sin\theta\sum_m \delta(t-m\tau)$$

$$\dot{\theta} = p_\theta / I$$

Between two consecutive impulses, p_θ is a constant and θ increases monotonically. After each impulse, the change of p_θ is:

$$\int_{t=n\tau+0^+}^{t=n\tau+1+0^+} dp_\theta = \int k \sin\theta \sum_m \delta(t - m\tau)dt = k \sin\theta_{n+1}$$

Hence, we have:

$$p_{n+1} = p_n + k \sin\theta_{n+1}$$

$$\theta_{n+1} = \theta_n + p_n \tau / I$$

(For convenience, we may let $\tau / I = 1$)

This is a very simple map and is often called a *standard map*. However, it possesses a very complicated dynamical structure. Fig 3.6 shows the cases with $k =$ 0.5, 1, 2 and 4. In the Figure, the closed curves are periodic or quasi-periodic trajectories (See Section 3.4). As the coupling strength k increases, as shown by the KAM theorem, these periodic, quasi-periodic trajectories will be destroyed gradually with the birth of more chaotic regions and new periodic, quasi-periodic trajectories (smaller rings in the Figure). As k is very large, the space is occupied mostly by chaotic trajectories. If a small portion of Fig.3.6 is enlarged, there will be more ring-like curves and chaotic regions, showing self-similarity.

The periodic and quasi-periodic trajectories shown in Fig.3.6 are very similar to those of a pendulum. In fact, the local dynamical structure is just that of a pendulum. This is a very important concept. We will address this in Chapter 9.

Furthermore, we note that the equations of motion for a pendulum are just the standard map. This is transparently clear if we note that the Hamiltonian of a pendulum is:

$$Gp^2 / 2 - F \cos\theta$$

and the equations of motion are:

$$dp / dt = -F \sin\theta$$

$$d\theta / dt = Gp$$

In terms of mapping, we have:

$$p_{n+1} - p_n = -F \sin\theta_n$$

$$\theta_{n+1} - \theta_n = Gp_{n+1}$$

This is identical to the standard map by rescaling.

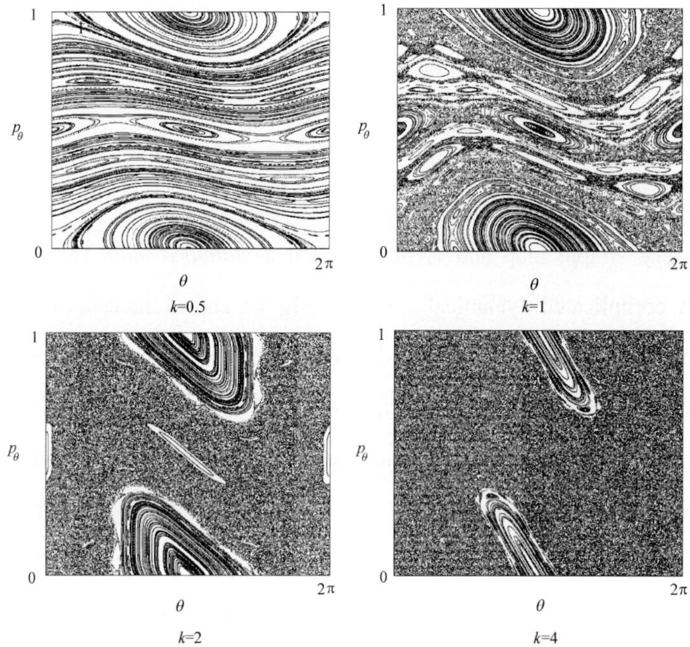

Fig.3.6 Standard map with various coupling strengths

3.7 Geometric and dynamical aspects of chaos

Dynamical chaos is different from randomness or commonly recognized disorder. Chaos is deterministic and can be represented by equations or maps. It may possess very complicated geometric structure, the so-called fractal, which is different from the common objects like point, segment, surface or body. A fractal can possess a fractional dimension. Meanwhile, chaos possesses a dynamical character which is realized by the Lyapunov exponent. A deep and significant topic is the relation between fractal dimensions and the Lyapunov exponent. Indeed, chaos has two aspects: the geometric and dynamical. We will study this topic in Chapters 11 and 13.

References

3.1 E. Ott, *Chaos in Dynamical Systems*, Cambridge University Press, Cambridge, 1994.

3.2 M.C. Gutzwiller, *Chaos in Classical and Quantum Mechanics in Interdisciplinary Applied Mathematics*, Vol. 1, Springer, Berlin, 1990.

Chapter 4

Application of su(2) algebra

4.1 Coupling of two Morse oscillators

The vibration of a chemical bond can be modeled as a Morse oscillator. Its energy is

$$\omega_0(n+\frac{1}{2})+\alpha(n+\frac{1}{2})^2$$

Here ω_0 is its fundamental frequency and α is the anharmonic coefficient (both are in the unit of cm^{-1}). In general, $\alpha<0$. As the quantum number (or action) n increases, the energy spacing becomes smaller. When the energy spacing between the nearest two levels is zero, the bond dissociates. The Morse oscillator is a good model for the vibration of a chemical bond. (For the discussion of Morse oscillator, see Section 1.2)

Two coupled identical Morse oscillators constitute a basic dynamical system. Its system Hamiltonian can be expressed as the sum of two identical Morse oscillators with an additional anharmonic term as:

$$\omega_0(n_1+n_2+1)+\alpha[(n_1+\frac{1}{2})^2+(n_2+\frac{1}{2})^2]+\alpha_{12}(n_1+\frac{1}{2})(n_2+\frac{1}{2})$$

and the 1:1 and 2:2（the so-called Darling–Dennison coupling）coupling (resonance) terms :

$$\frac{1}{2}[\beta+\frac{1}{2}\varepsilon(n_1+n_2+1)](a_1{}^+a_2+a_2{}^+a_1)$$

and

$$\delta(a_1{}^+a_1{}^+a_2a_2+a_2{}^+a_2{}^+a_1a_1)$$

Here, a^+, a are the creation and destruction operators, respectively. We can use the basis functions $|n_1\rangle|n_2\rangle$ $(\equiv|n_1,n_2\rangle)$ to construct the Hamiltonian matrix. By diagonalizing the matrix, the eigenfunctions and energies can be elucidated. The dimension of the matrix is $(n_1+1)\times(n_2+1)$. We note that $|n_1n_2\rangle$ are the eigenstates of the Hamiltonian without couplings. Under couplings, though individual n_1, n_2 are no longer good quantum numbers of the system nor are they conserved quantities, their sum n_1+n_2 is still conserved and is a good quantum number. This conserved quantity is called the polyad number. In fact, in the calculation, we may first assume a set of values for the coefficients, then by fitting the calculated eigenenergies to the experimental values (if they are available), the coefficients can be finally determined. The energy of molecular vibration (above the zero point energy) ranges from several hundred thousand wave numbers (cm^{-1}) to tens of thousand wave numbers. We have to expect that the fitted coefficients may vary in different energy ranges. Though the number of coefficients for fitting is multiple (say, up to 6 here), if the initial values are set reasonably, the fitting convergence is generally smooth and leads to unique final values by a nonlinear algorithm such as the Marquardt method.

One point that needs emphasis is that though the following analysis is more or less theoretical, the foundation is based on experiment, including the coefficients of the system algebraic Hamiltonian.

4.2 su(2) algebraic properties of a two-mode system

For two coupled identical Morse oscillators, we can define:

$$J_+ = a_1^+ a_2, \quad J_- = a_2^+ a_1, \quad J_z = \frac{1}{2}(a_1^+ a_1 - a_2^+ a_2)$$

The operators $\{J_+, J_-, J_z\}$ form an su(2) algebra. Its irreducible representations can be labeled by $J = \frac{1}{2}(n_1 + n_2)$. 1:1 and 2:2 couplings can be expressed by the elements of su(2) as:

$$\frac{1}{2}\beta'(J_+ + J_-) + \delta(J_+^2 + J_-^2)$$

with $\beta' \equiv \beta + \frac{1}{2}\varepsilon(2J+1)$

With these expressions under su(2), the system Hamiltonian is:

$$\omega_0(2J+1) + \frac{\beta'}{2}(J_+ + J_-) + (2\alpha + \alpha_{12})J(J+1) + (2\alpha + \alpha_{12})/4$$
$$+ (2\alpha - \alpha_{12})J_z^2 + \delta(J_+^2 + J_-^2)$$

Lie group theory and algebra show that a coset space SU(2)/U(1) can be constructed by the Lie group SU(2) and its subgroup U(1) (with subalgebra $\{J_z\}$). With this coset space, n_1, n_2, n_1^2, n_2^2 and $J_+ + J_-, J_+^2 + J_-^2$ can be expressed in terms of its coordinates (q, p). SU(2)/U(1) is a two dimensional sphere S^2 with radius J. Its spherical coordinates (θ, ϕ) are related to (q, p) via

$$q = \sqrt{J}\sin\frac{\theta}{2}\cos\phi\,, \quad p = -\sqrt{J}\sin\frac{\theta}{2}\sin\phi$$

The above procedure has been shown in Chapter 2. In this way, we have a Hamiltonian expressed in (q, p).

4.3 J_x, J_y, J_z as the coordinates of SU(2)/U(1) and the physical significance of the rotation of $\pi/2$ along J_y

In the coset representation, we have:

$$J_x \equiv \frac{1}{2}(J_+ + J_-) = 2q[J - (q^2 + p^2)]^{\frac{1}{2}} = J\sin\theta\cos\phi$$

$$J_y \equiv -\frac{i}{2}(J_+ - J_-) = -2p[J - (q^2 + p^2)]^{\frac{1}{2}} = J\sin\theta\sin\phi$$

$$J_z = 2(q^2 + p^2) - J = -J\cos\theta$$

It is apparent that $\{J_x, J_y, J_z\}$ can be the coordinates of the two dimensional sphere SU(2)/U(1) with radius J.

The rotation of $\frac{\pi}{2}$ along J_y possesses special significance. This is because the 1:1 resonance term $\beta' J_x$ will change to $\beta' J_z'$ in the new coordinates $\{J_x', J_y', J_z'\}$. In other words, the quantization changes from along J_z to $J_x(J_z')$. In the new coordinates, the 1:1 coupling term $\frac{\beta'}{2}J_z'$ possesses the eigenvector $|n_1' n_2'>$ with eigenvalue $\frac{\beta'}{2}(n_1' - n_2')$. Hence, we have the new Hamiltonian H' in the rotated coordinates as (Note that $J_x^{'2} = \frac{1}{2}[J(J+1) - J_z^{'2}] + \frac{1}{4}(J_+^{'2} + J_-^{'2})$)

$$\omega_0(2J+1) + \beta' J_z' + (3\alpha + \alpha_{12}/2)J(J+1) + (2\alpha + \alpha_{12})/4$$
$$-(\alpha - \alpha_{12}/2)J_z^{'2} + [\delta + (2\alpha - \alpha_{12})/4](J_+^{'2} + J_-^{'2})$$

Before rotation, the two Morse oscillators are identical, and we call the coordinates the local mode picture. After rotation, in H', the two Morse oscillators are no longer identical. We call the rotated coordinates the normal mode picture (In fact, n_1', n_2' are the actions of the symmetric and antisymmetric modes, respectively. We note also that $J = \frac{1}{2}(n_1 + n_2) = \frac{1}{2}(n_1' + n_2')$). If only the 1:1 coupling term is considered, the rotation of $\frac{\pi}{2}$ along the J_y axis is equivalent to the diagonalization of the Hamiltonian matrix H (since the 2:2 coupling term cannot be transformed to $J_z', J_z^{'2}$). This is equivalent to the normal mode analysis.

For these two pictures, we have (q, p) and (q', p'). Their relation can be determined by :

$$J_x' = -J_z, \quad J_y' = J_y, \quad J_z' = J_x$$

Or explicitly,

$$2q'[J - (q'^2 + p'^2)]^{\frac{1}{2}} = J - 2(q^2 + p^2) \tag{4.1}$$

$$-2p'[J - (q'^2 + p'^2)]^{\frac{1}{2}} = -2p[J - (q^2 + p^2)]^{\frac{1}{2}} \tag{4.2}$$

$$2(q'^2 + p'^2) - J = 2q[J - (q^2 + p^2)]^{\frac{1}{2}} \tag{4.3}$$

Dividing (4.2) by (4.1), we have

$$p' = \frac{2p[J - (q^2 + p^2)]^{\frac{1}{2}}}{J - 2(q^2 + p^2)} q' \tag{4.4}$$

By substituting (4.4) into (4.3), we have $q' = q'(q, p)$ Explicitly, it is:

$$q' = \text{sgn} \times [(q[J - (q^2 + p^2)]^{\frac{1}{2}} + \frac{J}{2})(1 + \frac{4p^2[J - (q^2 + p^2)]}{[J - 2(q^2 + p^2)]^2})^{-1}]^{\frac{1}{2}} \tag{4.5}$$

Here, $\text{sgn} = +1$, if $J - 2(q^2 + p^2) > 0$;

sgn = -1, if $J - 2(q^2 + p^2) < 0$;

as $J = 2(q^2 + p^2)$, $q' = p' = 0$

From (4.4) and (4.5), we have $p' = p'(q,p)$. (4.4) and (4.5) are very useful. As soon as the dynamics in one situation is known, then the dynamics in another situation can be known immediately via the transformation between (q,p) and (q',p').

4.4 Relation between Heisenberg's correspondence and the coset representation

Heisenberg recognized that for each quantum operator, there are corresponding classical variables: action n and angle φ, i.e.,

$$a_i^{+} \sim (n_i)^{\frac{1}{2}} e^{i\varphi_i}$$

$$a_i \sim (n_i)^{\frac{1}{2}} e^{-i\varphi_i}$$

From $J = \frac{1}{2}(n_1 + n_2)$, $J_z = \frac{1}{2}(n_1 - n_2)$ and $J_z = -J\cos\theta$, we have:

$$n_1^{\,2} = 4J^2 \sin^4(\frac{\theta}{2}),\quad n_2^{\,2} = 4J^2\cos^4(\frac{\theta}{2}),\quad n_1 n_2 = J^2 \sin^2\theta$$

$$\frac{1}{2}\beta'(a_1^{+}a_2 + a_2^{+}a_1) = \beta' J \sin\theta\cos\varphi$$

$$\delta(a_1^{+}a_1^{+}a_2a_2 + a_2^{+}a_2^{+}a_1a_1) = 2\delta J^2 \sin^2\theta\cos 2\varphi$$

Here, $\varphi = \varphi_1 - \varphi_2$ is the phase difference.

From $n_1^{\,2}(q,p)$, $n_2^{\,2}(q,p)$, $n_1n_2(q,p)$ and $\frac{1}{2}\beta'(a_1^{+}a_2 + a_2^{+}a_1)$, $\delta(a_1^{+}a_1^{+}a_2a_2 + a_2^{+}a_2^{+}a_1a_1)$ in terms of (q,p) (See Section (2.5)) and the relation between (q,p) and (θ,ϕ) of the S^2 sphere, we have:

$$n_1^{\,2} = 4J^2\sin^4(\frac{\theta}{2}) + 2J\sin^2(\frac{\theta}{2}),$$

$$n_2^{\,2} = 4J^2\cos^4(\frac{\theta}{2}) + \frac{1}{2}J\sin^2\theta,$$

$$n_1n_2 = J^2\sin^2\theta - \frac{1}{2}J\sin^2\theta$$

$$\frac{1}{2}\beta'(a_1^+a_2 + a_2^+a_1) = \beta' J\sin\theta\cos\phi$$

$$\delta(a_1^+a_1^+a_2a_2 + a_2^+a_2^+a_1a_1) = 2\delta J^2\sin^2\theta\cos 2\phi - \delta J\sin^2\theta\cos 2\phi$$

Comparing the results by Heisenberg's correspondence and coset representation, we know that as $J^2 >> J$, φ can be identified with ϕ. Then the results from Heisenberg's correspondence and the coset representation are quite close. The quantal effect by the coset representation is contained in the J term. For instance, as $n_1 + n_2 = 10$, $J = 5$, then $J/J^2 = \frac{1}{5}$ and the error is about 20%. The quantal effect is evident.

4.5 Dynamical representation of I_x and $I_+^2 + I_-^2$

From $\quad x_i \sim a_i^+ + a_i$

$$p_i \sim i(a_i^+ - a_i)$$

We have

$$x_1x_2 + p_1p_2 \sim a_1^+a_2 + a_2^+a_1 \sim I_x$$

Hence, I_x contains the couplings originating from both the potential coupling (depending on the position coordinates) and the momentum coupling.

Similarly, by neglecting the higher terms, we have:

$$I_+^2 + I_-^2 \sim (x_1x_2 + p_1p_2)^2$$

4.6 Dynamical analysis

From the analysis in Sections 4.3 and 4.4, we have $H(q,p)$ in the local mode picture and $H'(q',p')$ in the normal mode picture. The corresponding Hamilton's equations of motion are:

$$\partial H/\partial p = dq/dt,\quad \partial H/\partial q = -dp/dt$$

and

$$\partial H'/\partial p' = dq'/dt, \quad \partial H'/\partial q' = -dp'/dt$$

This is an integrable system. Its dynamics is known definitely by the given initial condition. Alternatively, we may equate the eigenenergy E to $H(q,p)$ or $H'(q',p')$ by which $q = q(p)$ or $q' = q'(p')$ can be obtained. In fact, for dynamical analysis, the relation (J_z, ϕ) or (J_z', ϕ') can be more useful. They show the relation of the action difference versus the angle difference in the local or normal mode picture.

In the following, we take H_2O and O_3 as examples. The coefficients for the Hamiltonians are: for H_2O, $\omega_0 = 3871.86$ cm^{-1}, $\alpha = -161.37$ cm^{-1}, $\alpha_{12} = -11.92$ cm^{-1}, $\beta = -111.77$ cm^{-1}, $\varepsilon = 10.98$ cm^{-1}, $\delta = -0.99$ cm^{-1} and for O_3, $\omega_0 = 1110.67$ cm^{-1}, $\alpha = -42.31$ cm^{-1}, $\alpha_{12} = 11.15$ cm^{-1}, $\beta = 49.12$ cm^{-1}, $\varepsilon = 11.52$ cm^{-1}, $\delta = -2.01$ cm^{-1}.

Given the total quantum number $N = n_1 + n_2$, there are $N+1$ levels. They are designated as L# with # = 1, 2, 3, …Levels are designated from the lowest to the highest.

There are two classes of levels as far as the dynamics is concerned. One is with two concentric trajectories in (q,p), which show up as two closed trajectories in (q',p'). In (J_z,ϕ) and (J_z',ϕ'), they are non-localized (ϕ extends from 0 to 2π) and localized (ϕ' is limited in a certain range), respectively. This is the character of the local mode for which the two oscillators are weakly coupled with no fixed phase difference. The two separated trajectories of the local mode in (J_z,ϕ) correspond to the clockwise and counterclockwise motions of the pendulum. Note that the trajectories of the local mode in the normal mode picture are 'localized' ! Its ϕ' is limited to 0 and π. Fig.4.1 shows the character of the local mode for the case of O_3, $N = 8$, L3 as an example.

Another class is with closed trajectories in both (q,p) and (q',p'). In (J_z,ϕ), ϕ is limited to around π and in (J_z',ϕ'), ϕ' is non-localized. This class is the normal mode. The localization of ϕ shows that the coupling between the two bonds is strong enough for them to possess a relatively fixed phase relation. That ϕ is around π shows that it is antisymmetric. (As the coupling is even stronger, for higher levels, ϕ can be limited to 0 and it is a symmetric normal mode.) Figure 4.1 shows the character of the normal mode for the case of H_2O, $N = 4$, L3 as an example.

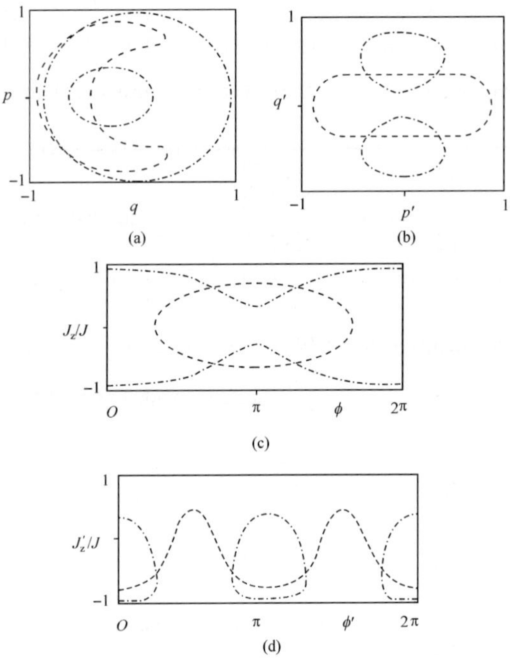

Fig.4.1 The trajectories of O_3, $N = 8$, L3 (______) and H_2O, $N = 4$, L3 (.......)

For a given N, there are N +1 levels. Those with lower energy are local modes while those with higher energy are normal. There is variation of the local/normal transition by Heisenberg's correspondence and coset algorithm for N <10. This shows that the quantal effect is significant for N <10. Tables 4.1 and 4.2 show the local (designated by L) and normal (designated by N) characters of the levels of H_2O and O_3 for various N, respectively. Those in parentheses show the local/normal transition by Heisenberg's correspondence. From the Tables, we note that the character of a level does not depend solely on its energy. For example, in H_2O, though L5 with N =

4 is lower than L1 with $N = 5$. The former is normal while the latter is local.

The dynamics by su(2) algebra for the two-mode system is integrable. The motion is periodic. Herein, there is no irregular or chaotic phenomenon which will show up in the highly-excited vibration of coupled three-mode system described by su(3) algebra.

Table 4.1 The characters of the levels of H_2O. Those in parentheses are by Heisenberg's correspondence. See text for details

Quantum number	level	energy（cm^{-1}）	local (L) /normal (N)	
$N=1$	L1	3654.12	N	
	L2	3754.91	N	
$N=2$	L1	7199.68	L	
	L2	7249.65	N	(L)
	L3	7443.12	N	(N)
$N=3$	L1	10601.60	L	
	L2	10615.70	L	(L)
	L3	10866.36	N	(N)
	L4	11031.88	N	
$N=4$	L1	13832.39	L	
	L2	13834.69	L	
	L3	14219.86	N	(L)
	L4	14319.27	N	(N)
	L5	14537.47	N	
$N=5$	L1	16895.43	L	
	L2	16895.70	L	
	L3	17458.92	N	
	L4	17495.54	N	(L)
	L5	17748.48	N	(N)
	L6	17948.08	N	

Table 4.2 The characters of the levels of O_3. Those in parentheses are by Heisenberg's correspondence. See text for details

Quantum number	level	energy (cm^{-1})	local (L) /normal (N)	
$N=6$	L1	5755.34	L	
	L2	5761.25	L	
	L3	5956.41	N	(L)
	L4	6044.40	N	(N)
	L5	6190.13	N	
	L6	6340.31	N	
	L7	6485.94	N	
$N=7$	L1	6572.44	L	
	L2	6574.89	L	
	L3	6826.94	N	
	L4	6892.91	N	(L)
	L5	7050.16	N	(N)
	L6	7206.83	N	
	L7	7373.06	N	
	L8	7528.78	N	
$N=8$	L1	7345.63	L	
	L2	7346.58	L	
	L3	7663.12	L	
	L4	7706.07	N	(L)
	L5	7882.38	N	(N)
	L6	8037.14	N	
	L7	8215.44	N	
	L8	8395.22	N	
	L9	8560.21	N	
$N=9$	L1	8075.54	L	
	L2	8075.90	L	
	L3	8458.12	L	
	L4	8482.14	L	(L)
	L5	8691.09	N	(N)
	L6	8833.52	N	
	L7	9020.58	N	
	L8	9213.91	N	
	L9	9406.73	N	
	L10	9580.18	N	
$N=10$	L1	8762.51	L	
	L2	8762.63	L	
	L3	9207.48	L	
	L4	9219.23	L	
	L5	9477.35	N	(L)
	L6	9596.86	N	(N)
	L7	9792.26	N	
	L8	9993.36	N	
	L9	10202.51	N	
	L10	10407.39	N	
	L11	10588.66	N	

References

4.1 G. Wu, *Chem. Phys. Lett.*, 209(1993)178

4.2 M. E. Kellman, *J. Chem. Phys.*, 83(1985)3843

Chapter 5

Application of noncompact su(1,1) algebra

5.1 Introduction

We established previously that for the vibrational two- and three-mode systems, su(2) and su(3) algebras can be applicable. In the application, conserved quantities are employed (corresponding to the compact S^2 and S^3 coset spheres) and the process is very neat. For the multiple modes, the algebras (su(n)) needed have a more complicated structure. This may hinder their application. In these cases, the direct product of noncompact su(1,1) algebra can be an alternative. The process will be more cumbersome than via the compact algorithm since a conserved quantity is not employed. However, its merit is that the algebraic structure is simply the direct product of the noncompact su(1,1) algebra and its application is methodologically simple. Here, the application of su(1,1) algebra will be demonstrated to show its characteristics and its superiority and inferiority to that of the su(2) algebra. (see Chapter 4)

5.2 Coset representatives of $SU(1,1)/U(1)_1 \otimes SU(1,1)/U(1)_2$ for two-mode system

The algebraic structure of SU(1,1) was shown in Section 2.5. It is composed of $\{K^+, K^-, K^0\}$ satisfying :

$$[K^0, K^\pm] = \pm K^\pm, \ [K^+, K^-] = -2K^0,$$

$$K^+|n\rangle = (n+1)|n+1\rangle$$

$$K^-|n\rangle = n|n-1\rangle$$

$$K^0|n\rangle = (n+\frac{1}{2})|n\rangle$$

Here, $|n\rangle$ is an eigenfunction, n = 0, 1, 2, ...(k = 1/2 is chosen and its labeling is neglected as stated in Section 2.5). The coset space is SU(1,1)/U(1) (u(1) algebra is K^0) with noncompact hyperbolic surface. Its coordinates can be (q, p) for which:

$$\frac{1}{\sqrt{2}}(q+ip) = \sinh(\theta/2)e^{-i\phi}$$

Here, $-\infty < \theta < \infty$, $0 \le \phi \le 2\pi$ are the polar coordinates of the hyperbolic surface. We have the following representations for these operators (see Section 2.5 for the details.).

$$K^0 = \frac{1}{2}(q^2 + p^2 + 1)$$
$$K^- = \frac{1}{2}(q+ip)[q^2 + p^2 + 2]^{1/2}$$
$$K^+ = \frac{1}{2}(q-ip)[q^2 + p^2 + 2]^{1/2}$$

The algebraic Hamiltonian of a two-mode system is then:

$$\omega_1(n_1 + 1/2) + \omega_2(n_2 + 1/2) + \alpha_1(n_1 + 1/2)^2 + \alpha_2(n_2 + 1/2)^2$$
$$+\alpha_{12}(n_1 + 1/2)(n_2 + 1/2) + \beta(a_1^+ a_2 + a_2^+ a_1)$$

In (q_1, p_1, q_2, p_2), it is

$$\omega_1/2(q_1^2 + p_1^2 + 1) + \omega_2/2(q_2^2 + p_2^2 + 1) + \alpha_1/4(q_1^2 + p_1^2 + 1)^2 + \alpha_2/4(q_2^2 + p_2^2 + 1)^2$$
$$+\alpha_{12}/4(q_1^2 + p_1^2 + 1)(q_2^2 + p_2^2 + 1)$$
$$+\beta/2(q_1 q_2 + p_1 p_2)[(q_1^2 + p_1^2 + 2)(q_2^2 + p_2^2 + 2)]^{1/2}$$

Here, ω_1 and ω_2 are the mode frequencies, α_1, α_2, α_{12} are the anharmonic coefficients and the β term is the coupling between the two modes which can be represented in terms of the operators of su(1, 1) as :

$$\beta(K_1^+ K_2^- + K_1^- K_2^+)$$

In (q_1, p_1, q_2, p_2), Hamilton's equations of motion are:

$$\frac{\partial H(q,p)}{\partial p_i} = \frac{dq_i}{dt}, \quad \frac{\partial H(q,p)}{\partial q_i} = -\frac{dp_i}{dt} \quad (i = 1, 2)$$

and the dynamics can be described in the $SU(1,1)/U(1)_1 \otimes SU(1,1)/U(1)_2$ coset space with the coordinates (q_1, p_1, q_2, p_2).

5.3 Contrast of su(1,1) to su(2)

In the last section, the coefficients of the Hamiltonian, ω_1, ω_2, α_1, α_2 and α_{12} (except β) are the same as those in the su(2) representation. (See Chapter 4). This can be checked by letting $n_2 = N - n_1$ with N the conserved quantity. However, the β term of the 1:1 coupling is not as compact as that in the su(2) represented Hamiltonian:

$$\beta' q/2[2N-(q^2+p^2)]^{1/2}$$

This is because in the noncompact su(1,1) representation, the conserved N is not adequately exploited. Besides, we note that β and β' terms in q_i, p_i (i=1,2) are of orders 4 and 2, respectively.

It is obvious that by the noncompact algorithm, the extension to the general n-mode case is straightforward by the direct product of n SU(1,1)/U(1) coset spaces (simply denoted as $\otimes$ SU(1,1)/U(1)^n). For the compact algorithm, the extension from SU(2)/U(1) to SU(n)/U(n-1) is cumbersome. For comparison, in $\otimes$ SU(1,1)/U(1)^n representation, there are $2n$ Hamilton's equations while in SU(n)/U(n-1) representation, there are 2(n -1) equations. The orders of the 1:1 coupling term in q_i, p_i are 4 and 2, respectively. As n is larger, the advantage of the compact algorithm is not so obvious.

Another advantage of the noncompact algorithm is its easiness in treating the general coupling forms among the vibrational modes. For example, the 2:3 coupling $a_1^+a_1^+a_2a_2a_2$ in the two-mode system can be easily expressed as a function of (q_i, p_i) in the noncompact algorithm. This is not so straightforward in the compact SU(2)/U(1) space.

We have to learn that though in $\otimes$ SU(1,1)/U(1)^2 no constants are employed, there is indeed such a conserved quantity as can be checked from the following consideration. The variations of n_1 and n_2 as the function of time are :

$$\frac{d}{dt}n_1 = \beta/2(q_1p_2 - q_2p_1)[(q_1^2 + p_1^2 + 2)(q_2^2 + p_2^2 + 2)]^{1/2}$$

and

$$\frac{d}{dt}n_2 = \beta/2(-q_1p_2 + q_2p_1)[(q_1^2 + p_1^2 + 2)(q_2^2 + p_2^2 + 2)]^{1/2}$$

Hence,

$$\frac{d}{dt}N = \frac{d}{dt}(n_1 + n_2) = 0$$

In other words, *N* is conserved.

In the su(2) representation, the 1:1 coupling term is $\beta'/2(J_+ + J_-)$. It shows the destruction of the action of mode 2 (1) and the creation of the action of mode 1 (2). $a_1^+a_2 + h.c.$ appears as an integral J_+ (J_-) operator. In the noncompact su(1,1) representation, the 1:1 coupling is $\beta(K_1^+K_2^- + h.c.)$ in which the coupling is expressed as two *independent* events. The difference lies in that the former is of order 2 while the latter is of order 4. Hence, β will be smaller β'. We estimate that β should not be less than $2\beta'/N$. The exact β has to be determined by the fit of the eigenenergies of the noncompact Hamiltonian to the experimental values.

In su(2) representation, the system is integrable. In (*q, p*), the normal and local modes correspond to a single and two closed curves, respectively. The phase angle of the normal mode is limited to a certain range while that of the local mode can extend in the full range of $(-\pi, \pi)$. (This is in the local mode picture. See Chapter 4.) In the su(1,1) representation, we have a four dimensional (q_1, p_1, q_2, p_2) space. The two trajectories in (q_1, p_1) and (q_2, p_2), respectively, are no longer closed and quite possibly that they are quasi-periodic. Clearly, as α_1, α_2, α_{12} and β are zero, and if ω_1, ω_2 are not in rational ratio, the trajectory is quasi-periodic. Due to weak perturbation (by the small α_1, α_2, α_{12} and β), quasi-periodicity may not be destroyed. What is certain is that there is correlation between the motions in (q_1, p_1) and (q_2, p_2). This justifies that the closure of the trajectories in (q, p) coordinates in the su(2) representation should bear its expression in the noncompact (q_1, p_1) and (q_2, p_2) spaces. This will be demonstrated in the following numerical simulation.

5.4 Numerical simulation

In the compact su(2) representation, the Hamilton's equations of motion are related to $\omega_1 - \omega_2$, while in the noncompact su(1, 1) representation, they are related to both ω_1 and ω_2. Their orders of the Hamilton's equations are 2 and 4, respectively. In numerical calculation for reaching comparative degree of accuracy, in the noncompact case, the accuracy should be of two more orders than in the compact case. Therefore, the time consumption for the noncompact case will be more than a hundred fold. This is a serious disadvantage of the noncompact algorithm.

The data for the stretching and bending of CH of (CF_3) CH (Ref.5.2) are adopted. They are ω_1 = 3042.8 cm^{-1}, ω_2 = 1356.4 cm^{-1}, α_1 = 56.9 cm^{-1}, α_2 = −0.32 cm^{-1}, α_{12} = −14.0 cm^{-1}. β is taken as −10.0 cm^{-1}. Initial $(q_1, p_1, q_2, p_2)_0$ = (−1.0000, 3.0000, 4.4306, 3.2202), n_1^0=5, n_2^0=15. The integration step is 0.33×10^{-3} fs for a duration of 5.045 ps. In Fig.5.1 (a), the inner and outer trajectories are in the (q_1, p_1) and (q_2, p_2) spaces, respectively. The time behavior of n_1 and n_2 is shown in Fig. 5.1 (b). It shows the action (energy) transfer between the two modes. Fig. 5.1 (c) shows the relation between $n_1 - n_2$ and the phase angle ϕ. ϕ is $\tan^{-1}(-p_1/q_1) - \tan^{-1}(-p_2/q_2)$. The distinct single curve shows that the motions in (q_1, p_1) and (q_2, p_2) spaces are well correlated. This is consistent with the result by the su(2) algorithm. Since ϕ is extensive in between $-\pi$ and π, the motion is a local mode. As the coupling is stronger, the normal mode may appear as shown in

Fig.5.1 (d) (The data employed are $n_1^0 = 5$, $n_2^0 = 15$, $\omega_1 = \omega_2 = 3042.8\text{cm}^{-1}$, $\alpha_1 = \alpha_2 = -56.9\ \text{cm}^{-1}$, $\alpha_{12} = -14.0\ \text{cm}^{-1}$ and $\beta = -60.0\ \text{cm}^{-1}$). This is the same as by the su(2) algorithm.

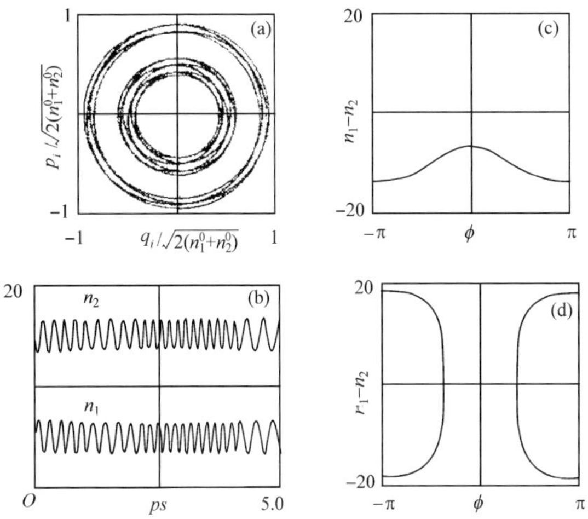

Fig.5.1(a) The inner and outer trajectories are in the (q_1, p_1) and (q_2, p_2) spaces respectively; (b) the time behavior of n_1 and n_2; (c) the relation between $n_1 - n_2$ and phase angle ϕ; and (d) the case of the normal mode. See text for details

In summary, the main shortcomings for the noncompact algorithm are the increase of the number of the Hamilton's equations and a hundred-fold increase of computational time as a consequence of the increase of its equation orders. However, the advantage of the noncompact algorithm is also obvious, namely its convenience.

References

5.1 G. Wu, *Chem.Phys.Lett.*,248(1996)77.

5.2 J. E. Baggott, M. C. Chuang, R. N. Zare, H. R. Dubal and M. Quack, *J. Chem. Phys.*, 82(1985)1186.

Chapter 6

Breaking of su(3) algebra and its application

6.1 Breaking of su(3) algebra

The three-mode system possesses a much more complicated dynamics than the two-mode system. The dynamics of the two-mode system is integrable under 1:1 coupling except that when more couplings such as 2:3 resonance are present it may become unintegrable. A three-mode system is in general not integrable due to couplings and its dynamics is very complicated and includes periodic, quasi-periodic, irregular and chaotic motions. In the previous chapters, we have introduced the Lie algebraic algorithm (See Refs.6.1 6.5) to two-mode systems. The extension to three-mode system is straightforward. However, the dynamics will be drastically different. In this chapter, we will pay attention to its algebraic properties and dynamical behavior to lay the foundation of the algebraic approach to the molecular highly excited vibration.

For the three-mode system, we have the creation and destruction operators a_i^+, a_i $(i = 1, 2, 3)$ to describe its dynamics. Their actions are:

$$n_i = a_i^+ a_i$$

and the commutators are:

$$[a_i, a_j^+] = \delta_{ij}, \quad [a_i, a_j] = [a_i^+, a_j^+] = 0$$

Here δ_{ij} is 0 as $i \neq j$ and is 1 as $i = j$.

Meanwhile, we may define the following operators:

$$I_+ = a_1^+ a_2, \quad J_+ = a_1^+ a_3, \quad K_+ = a_2^+ a_3,$$

$$I_z = (n_1 - n_2)/2, \quad Y = (n_1 + n_2 - 2n_3)/2,$$

$$I_- = (I_+)^+, \quad J_- = (J_+)^+, \quad K_- = (K_+)^+$$

These operators satisfy the su(3) algebraic structure. In other words, there are:

$$[I_+, I_-] = 2I_z\,,\quad [I_z, I_\pm] = \pm I_\pm\,,\quad [I_z, J_\pm] = \pm J_\pm/2\,,$$

$$[I_z, K_\pm] = \mp K_\pm/2\,,\quad [Y, J_\pm] = \pm 3J_\pm/2\,,\quad [Y, K_\pm] = \pm 3K_\pm/2\,,$$

$$[J_+, J_-] = I_z + Y\,,\quad [K_+, K_-] = Y - I_z\,,\quad [J_\pm, K_\pm] = 0$$

From the above relations, it is obvious that $\{I_z, I_+, I_-\}$ act as the angular momentum operators and form an su(2) algebra, which is a subalgebra of su(3).

For a three-mode system, its Hamiltonian up to the second order anharmonicities without couplings can be written as:

$$\begin{aligned}
&\omega(n_1 + n_2 + n_3 + \frac{3}{2}) \\
&+\alpha[(n_1 + \frac{1}{2})^2 + (n_2 + \frac{1}{2})^2 + (n_3 + \frac{1}{2})^2] \\
&+\alpha'[(n_1 + \frac{1}{2})(n_2 + \frac{1}{2}) + (n_1 + \frac{1}{2})(n_3 + \frac{1}{2}) + (n_2 + \frac{1}{2})(n_3 + \frac{1}{2})] \qquad (6.1)
\end{aligned}$$

Here, ω is the frequency. α , α' are anharmonic coefficients. All the units are cm^{-1}.

As stated before, the dynamics of a three-mode system is very complicated. As a first step, we consider the following two couplings:

Case A: $t(a_1^+ a_2^+ a_3 a_3 + h.c.)$

Case B: $t(a_1^+ a_2 + h.c.)$

In case A, I_z is conserved while in case B, Y is conserved. For these couplings, they are the conserved quantities. This will help in exploring their dynamics.

With the su(3) operators, (6.1) can be written as:

$$\begin{aligned}
&\omega(3I + \frac{3}{2}) + 2\alpha[\frac{3}{2}I(I+1) + \frac{1}{3}Y^2 + I_z^{\,2} + \frac{8}{3}] \\
&+\alpha'[3I(I+1) - \frac{1}{3}Y^2 - I_z^{\,2} + \frac{3}{4}] \qquad (6.2)
\end{aligned}$$

For the couplings, Heisenberg's correspondence for the second quantized operators can be adopted for a classical treatment:

$$a^+ \sim (n + \frac{1}{2})^{\frac{1}{2}} e^{i\phi}$$

$$a \sim (n + \frac{1}{2})^{\frac{1}{2}} e^{-i\phi}$$

Note that (n,ϕ) form the classical action-angle variables.

Therefore, we have:

Case A: $2t\{[(n_1+\frac{1}{2})(n_2+\frac{1}{2})]^{\frac{1}{2}}(n_3+\frac{1}{2})\cos\varphi\}$

Case B: $2t\{[(n_1+\frac{1}{2})(n_2+\frac{1}{2})]^{\frac{1}{2}}\cos\phi$

with $\varphi=\phi_1+\phi_2-2\phi_3$ and $\phi=\phi_1-\phi_2$. Their associated actions are Y and I_z, respectively.

In terms of the su(3) algebraic operators, the above expressions can be cast to:

Case A: $2t\{[(I+\frac{1}{3})^2-I_z^{\ 2}+(I+\frac{1}{3}Y)+\frac{1}{4}]^{\frac{1}{2}}(I-\frac{2}{3}Y+\frac{1}{2})\cos\varphi\}$

Case B: $2t\{[(I+\frac{1}{3})^2-I_z^{\ 2}+(I+\frac{1}{3}Y)+\frac{1}{4}]^{\frac{1}{2}}\cos\phi$ (6.3)

The system Hamiltonian is composed of (6.2) and (6.3). Since $I=(n_1+n_2+n_3)/3$ is conserved, the dynamics can be fully described by $\{Y,\varphi\}$ and $\{I_z,\phi\}$. This leads to the description of dynamics by trajectories. The Hamilton's equations of motion for $\{Y,\varphi\}$ and $\{I_z,\phi\}$ are:

$$\dot{Y}=-\partial H/\partial\varphi,\quad \dot{\varphi}=\partial H/\partial Y$$

and

$$\dot{I}_z=-\partial H/\partial\phi,\quad \dot{\phi}=\partial H/\partial I_z$$

Or explicitly as: (Note that in Cases A and B, $\dot{I}_z$ and $\dot{Y}$ are 0, respectively.)

Case A:

$$\dot{Y}=2t[(I+\frac{1}{3}Y)^2-I_z^{\ 2}+(I+\frac{1}{3}Y)+\frac{1}{4}]^{\frac{1}{2}}(I-\frac{2}{3}Y+\frac{1}{2})\sin\varphi$$

$$\dot{\varphi}=\frac{2}{3}(2\alpha-\alpha')Y+\frac{2}{3}t\{[(I+\frac{1}{3}Y)^2+(I+\frac{1}{3}Y)+\frac{1}{4}-I_z^{\ 2}]^{-\frac{1}{2}}(I-\frac{2}{3}Y+\frac{1}{2})$$

$$\times(I+\frac{1}{3}Y+\frac{1}{2})-2[(I+\frac{1}{3}Y)^2+(I+\frac{1}{3}Y)+\frac{1}{4}-I_z^{\ 2}]^{\frac{1}{2}}\}\cos\varphi$$

$$\dot{\phi}=2(2\alpha-\alpha')I_z-2t[(I+\frac{1}{3}Y)^2+(I+\frac{1}{3}Y)+\frac{1}{4}-I_z^{\ 2}]^{-\frac{1}{2}}(I-\frac{2}{3}Y+\frac{1}{2})I_z\cos\varphi$$

Case B:

$$\dot{I}_z=2t[(I+\frac{1}{3}Y)^2+(I+\frac{1}{3}Y)+\frac{1}{4}-I_z^{\ 2}]^{\frac{1}{2}}\sin\phi$$

$$\dot{\phi}=2(2\alpha-\alpha')I_z-2t[(I+\frac{1}{3}Y)^2+(I+\frac{1}{3}Y)+\frac{1}{4}-I_z^{\ 2}]^{-\frac{1}{2}}I_z\cos\phi$$

$$\dot{\varphi}=\frac{2}{3}(2\alpha-\alpha')Y+\frac{2}{3}t[(I+\frac{1}{3}Y)^2+(I+\frac{1}{3}Y)+\frac{1}{4}-I_z^{\ 2}]^{-\frac{1}{2}}(I+\frac{1}{3}Y+\frac{1}{2})\cos\phi$$

As the system Hamiltonian is without couplings as shown in (6.2), n_1, n_2 and n_3 are conserved. So are I_z and Y. Under the couplings, Y and I_z are not conserved, though there is still the conserved quantity $I=(n_1+n_2+n_3)/3$. Therefore, we say that su(3) algebra is broken. Besides, we note that (6.2) is symmetric under

$$Y\to -Y,\ I_z\to -I_z$$

and the coupling term (6.3), for both Case (A) and (B), possesses the symmetry with respect to

$$I_z\to -I_z$$

Hence, the trajectories are symmetric under $I_z\to -I_z$. Meanwhile for Case (A), the eigenstates with the same $|I_z|\neq 0$ are degenerate. As $I>>Y$, (6.3) is symmetric under $Y\to -Y$. In this situation, the system may possess approximate constants of motion. This will be one of the properties of interest for the highly excited vibration of which I is large.

6.2 Numerical simulation

Suppose that the coefficients of the system Hamiltonian are known, by employing the basis functions $|n_1>|n_2>|n_3>$, then the eigenfunctions and energies can be obtained. Conversely, if the eigenenergies are known by experiment, together with the functional form of the system Hamiltonian (especially the coupling forms), then by fitting, the coefficients of the Hamiltonian can be determined.

For an eigenenergy E, we may find the solutions to $E = H(Y, \varphi, I_z = \text{constant})$ (Case A) or $E = H(I_z, \phi, Y = \text{constant})$ (Case B) to obtain $Y = Y(\varphi)$ or $I_z = I_z(\phi)$. These relations realize the dynamical properties of the system. Of course, by the integration of the Hamilton's equations, the time evolution of Y, φ, ϕ or I_z, ϕ, φ can be obtained.

In solving these equations, we have to limit Y and I_z to the following ranges, respectively:

Case A: $-3(I-|I_z|) \le Y \le \frac{3}{2}I$

Case B: $-(I+\frac{1}{3}Y) \le I_z \le I+\frac{1}{3}Y$

These constraints are deduced from the very definitions of I_z and Y as functions of n_1, n_2 and n_3.

For numerical calculation, α, α' and ω are taken as -50.45cm^{-1}, 0.15 cm^{-1} and 3428.99 cm^{-1} (mimic of the C–H stretch. See Ref.6.1) t is taken as 1000 cm^{-1}, 100 cm^{-1} and 10 cm^{-1} to show strong, medium and weak couplings. In addition, $I = \frac{7}{3}$.

Figures. 6.1 and 6.2 show the calculated levels. In Fig. 6.1, the levels are degenerate when $|I_z| \ne 0$. For strong coupling, both Cases A and B show rather regular patterns similar to the staircase except those marked by * in Fig. 6.1(c). Meanwhile, the increments of $|I_z|$, Y among the nearest two levels are $\frac{1}{2}$ and $\frac{3}{2}$, respectively. The level spacings are almost equidistant, except for those in the low and high levels as shown in Fig. 6.1(c) for Case A for which the level spacings are less. All these show the normal character. For the weak coupling of Case B, the levels are almost degenerate as shown in Fig. 6.2(a). These demonstrate that the three oscillators are rather independent. Though the dynamics of highly excited vibration is complicated, there is indeed structure.

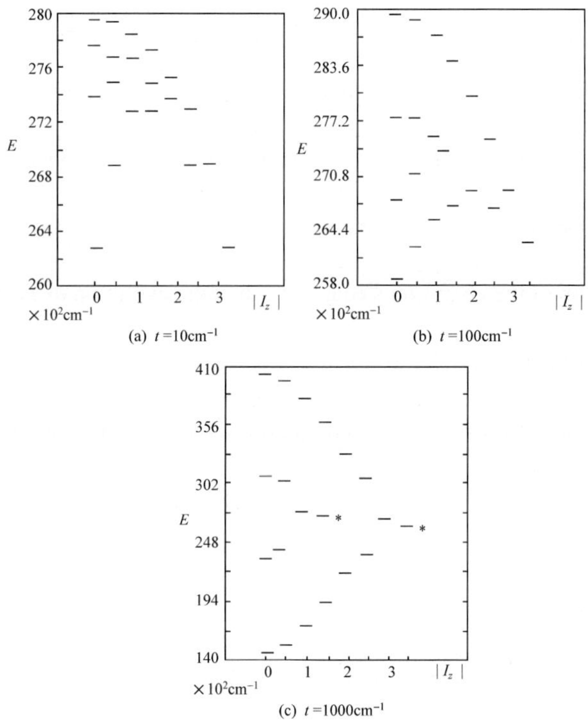

Fig.6.1 The energy levels for Case A, $I = 7/3$

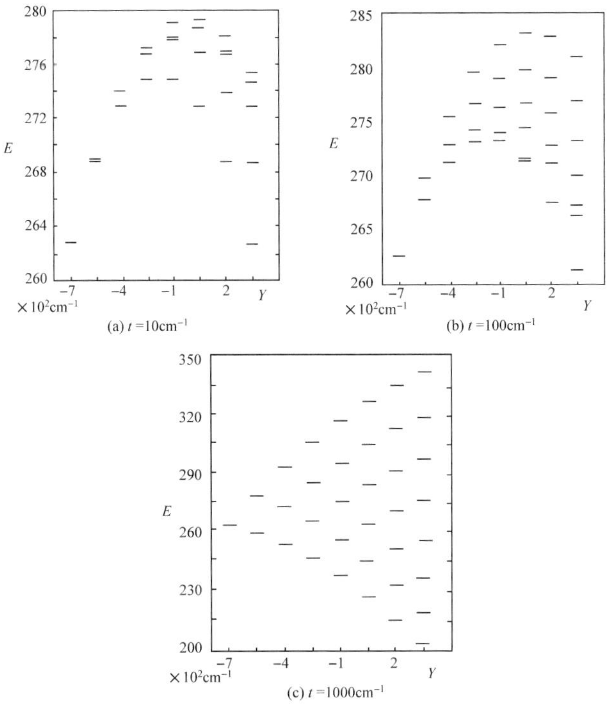

Fig.6.2 The energy levels for Case B, $I = 7/3$

In the following, we will discuss the dynamical properties under weak, medium and strong couplings in Cases A and B.

6.2.1 Case A

(1) Weak coupling

In $Y(\varphi)$, φ for the low levels will be extensive in $[0,2\pi]$ while for the high levels φ is limited to $(-\frac{\pi}{2},\frac{\pi}{2})$. For a demonstration, Fig.6.3(a),(b) show the characters of these two trajectories. For delocalized φ, as $|I_z|$ is large, Y is larger than 0 (I_z is constant) while as $|I_z|$ is smaller, Y is negative, in general. The former situation corresponds to a smaller n_3 while the latter to a larger n_3. This shows that the action of the third oscillator is relevant to the action difference between the other two.

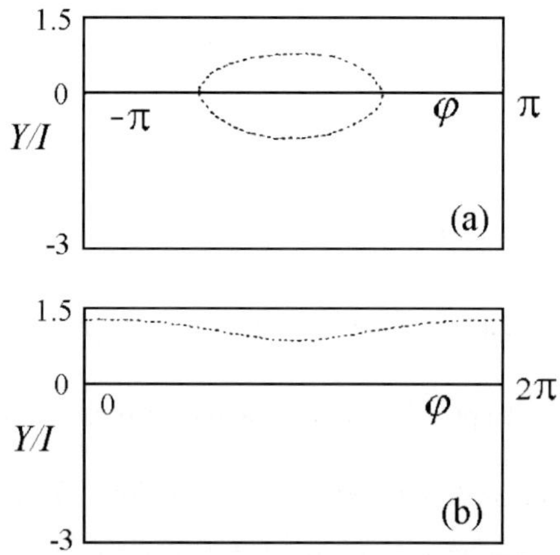

Fig.6.3 Trajectories for Case A, $I = 7/3$, $t = 10\ \text{cm}^{-1}$ (a) $E = 27937\ \text{cm}^{-1}$ $|I_z| = 0.5$

(b) $E = 27476\ \text{cm}^{-1}$ $|I_z| = 1.5$

(2) Medium coupling

For most levels, φ is limited to $(-\frac{\pi}{4}, \frac{\pi}{4})$, showing that the coupling among the oscillators is strong enough for their phase angles to be correlated in a certain range. An interesting situation is a bell-like trajectory as shown in Fig. 6.4 ($I_z = 0$) in which Y is oscillating between $1.5I$ and $-3I$.

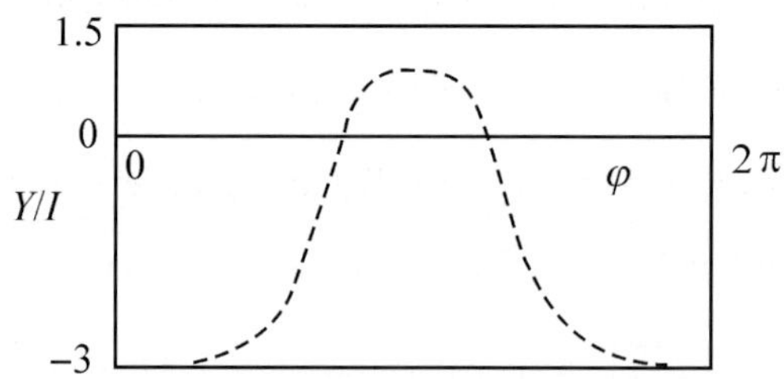

Fig.6.4 Trajectories for Case A, $I = 7/3$, $t = 100\ \text{cm}^{-1}$, $E = 26798\ \text{cm}^{-1}$, $|I_z| = 0$

(3) Strong coupling

In this situation, the phase angles for all the levels are localized. For low levels, φ is limited to $(\frac{3\pi}{4},\frac{5\pi}{4})$ while for the high levels, φ is limited to $(-\frac{\pi}{4},\frac{\pi}{4})$.

6.2.2 Case B

In Case B, $I_z(\phi)$ possesses the symmetry under the transformation $I_z \to -I_z$. Trajectories can be classified into two classes: one with ϕ extensive in $(0,2\pi)$ and the other with ϕ limited in a specific range. For the former, the coupling between oscillators 1 and 2 is weak with no fixed phase relation while the latter shows 'phase-locking' phenomenon. They correspond to local and normal modes, respectively. In the local mode, the action transferred between the two oscillators is less, and there are two corresponding trajectories.

(1) Weak coupling

In this situation, both local and normal modes are evident. In the local mode, lower levels show less variation of I_z against ϕ, indicating less action (energy) transferred. The mode character (local or normal) of a level depends not only on its energy but also on Y. Fig.6.5(a),(b) show the trajectories of two levels sharing the same $Y = 0.5$ where the energy difference is $58\,cm^{-1}$. The higher level is normal and the lower one is local as expected. Fig. 6.6(a),(b) show the trajectories of two levels with an energy difference of $18\,cm^{-1}$ (of the same order as the coupling energy) but their Y values are 2 and -2.5, respectively. The slightly higher level is normal and the lower one is local.

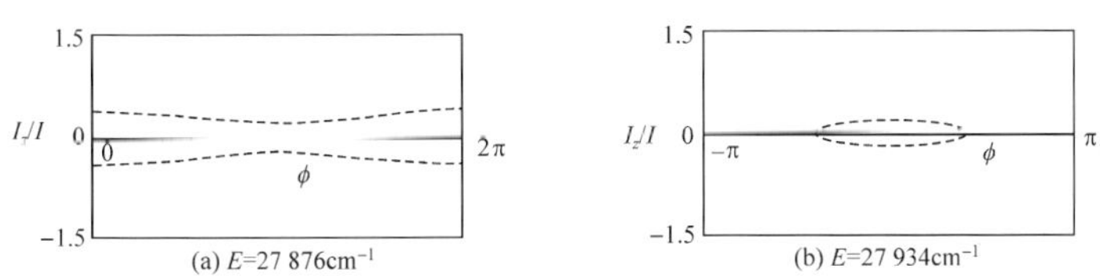

Fig.6.5 Trajectories for Case B, $I = 7/3$, $t = 10$ cm^{-1}, Y=0.5

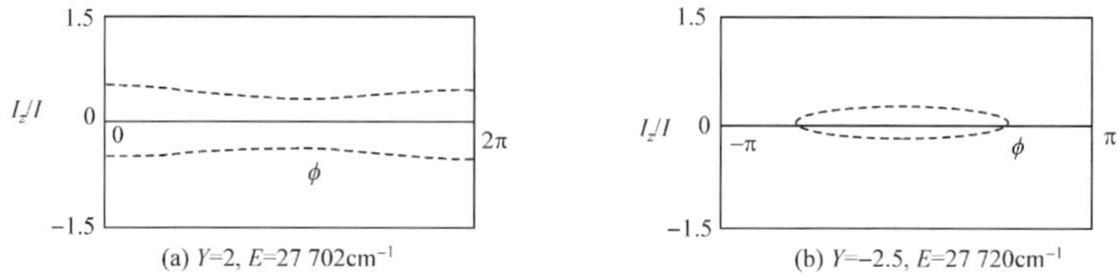

Fig.6.6 Trajectories for Case B, $I = 7/3$, $t = 100$ cm^{-1},

(2) Medium coupling

Under such a coupling strength, most high levels are normal with ϕ limited to $(-\frac{\pi}{2}, \frac{\pi}{2})$. Similarly, both energy E and Y have an effect on the mode character. For instance, two levels of energy $E = 27339$ cm^{-1} and 27037 cm^{-1} share the same $Y = 3.5$. Their energy difference is 302 cm^{-1}. The higher one is normal and the lower one is local as anticipated. Besides, we note that although the level of E = 26991 cm^{-1} is lower than that of 27037 cm^{-1}（the difference is only 46 cm^{-1}, less than the coupling strength）its $Y = -5.5$ and is of normal character.

(3) Strong coupling

In this situation, all the levels are of normal mode. For the lower levels, ϕ is limited to around π while for the higher ones, ϕ is around 0. The former corresponds to the antisymmetric mode with respect to the oscillators 1 and 2 while the latter is symmetric. The situation is analogous to the two O–H stretches where there are antisymmetric and symmetric modes. The antisymmetric mode is of lower energy.

For the general coupled three-mode systems, the breaking of su(3) algebra is very serious. There are no conserved quantities except energy. It is not integrable and a full understanding of its dynamics is difficult. However, as the coupling is of a certain special form such that some constants of motion just like I_z, Y and I exist, a thorough exploration is possible. Fermi resonance is such a case, as shown in the next section.

6.3 su(3) represented Fermi resonance

For a stretch motion s, if its frequency is twice that of its bend b, then there can be Fermi resonance among them. From the creation and destruction operators of s and b, the following new operators can be formed:

$$I_+^{'} = a_s^+ a_b a_b / \sqrt{a_b^+ a_b} \,, \quad I_-^{'} = a_b^+ a_b^+ a_s / \sqrt{a_b^+ a_b}$$

$$I_z^{'} = \frac{1}{2}(a_s^+ a_s - \frac{1}{2} a_b^+ a_b), \quad I^{'} = \frac{1}{2}(a_s^+ a_s + \frac{1}{2} a_b^+ a_b)$$

It can be shown that $\{I_+^{'}, I_-^{'}, I_z^{'}\}$ form an su(2) algebra (Ref.6.6).

We may extend this inference to the case of which two oscillators s and t that are coupled via 1:1 resonance are in Fermi resonance with their bend. Define:

$$I_+ = a_s^+ a_t \,, \quad J_+ = a_s^+ a_b a_b / \sqrt{a_b^+ a_b} \,,$$

$$K_+ - a_t^+ a_b a_b / \sqrt{a_b^+ a_b} \,, \quad I_z = \frac{1}{2}(a_s^+ a_s - a_t^+ a_t)$$

$$Y = \frac{1}{2}(a_s^+ a_s + a_t^+ a_t - a_b^+ a_b) \,, \quad I = \frac{1}{3}(a_s^+ a_s + a_t^+ a_t + \frac{1}{2} a_b^+ a_b)$$

$$I_- = (I_+)^+ \,, \quad J_- = (J_+)^+ \,, \quad K_- = (K_+)^+$$

By direct calculation, it can be shown that $\{I_+, I_-, I_z\}$ form an su(2) algebra, i.e.,

$$[I_+, I_-] = 2I_z \,, \quad [I_z, I_\pm] = \pm I_\pm$$

Furthermore, we have

$$[I_z, J_\pm] = \pm\frac{1}{2} J_\pm \,, \quad [I_z, K_\pm] = \mp\frac{1}{2} K_\pm$$

$$[J_\pm, K_\pm] = 0$$

By

$$[I_+^{'}, I_-^{'}] = 2I_z^{'}$$

then

$$[K_+, K_-] = \frac{1}{2}(a_t^+ a_t - a_b^+ a_b) = Y - I_z$$

Analogously,

$$[J_+, J_-] = I_z + Y,$$

Besides, from

$$[I_z', I_\pm'] = \pm I_\pm'$$

we have

$$[a_b^+ a_b, a_s^+ a_b a_b / \sqrt{a_b^+ a_b}] = -2 a_s^+ a_b a_b / \sqrt{a_b^+ a_b},$$

From this, it can be shown that:

$$[Y, K_+] = \frac{3}{2} K_+$$

Analogously,

$$[Y, K_-] = -\frac{3}{2} K_-$$

$$[Y, J_\pm] = \pm\frac{3}{2} J_\pm$$

Hence, $\{I_\pm, J_\pm, K_\pm, I_z, Y\}$ form an su(3) algebra.

For a system with two identical *s*, *t* stretches and one *b* bend oscillators, its Hamiltonian can be written as:

$$H = H_0 + H'$$

$$H_0 = \omega_s (n_s + n_t + 1) + \omega_b (n_b + 1) + X_{ss}[(n_s + \frac{1}{2})^2 + (n_t + \frac{1}{2})^2]$$

$$+ X_{bb}(n_b + 1)^2 + X_{st}(n_s + \frac{1}{2})(n_t + \frac{1}{2}) + X_{sb}(n_s + n_t + 1)(n_b + 1)$$

$$H' = K_{sbb}(a_s^+ a_b a_b + a_t^+ a_b a_b + h.c.) + K_{st}(a_s^+ a_t + h.c.) \qquad (6.4)$$

The K_{st} term in H' shows the 1:1 coupling between *s* and *t* oscillators. The K_{sbb} term is for Fermi resonance.

Through the su(3) operators, H_0, H' can be written as

$$H_0 = \omega_s (2I + \frac{2}{3} Y + 1) + \omega_b (2I - \frac{4}{3} Y + 1)$$

$$+ X_{ss}[(I + \frac{1}{3} Y + I_z + \frac{1}{2})^2 + (I + \frac{1}{3} Y - I_z + \frac{1}{2})^2]$$

$$+ X_{bb}(2I - \frac{4}{3} Y + 1)^2$$

$$+X_{st}(I+\frac{1}{3}Y+I_z+\frac{1}{2})(I+\frac{1}{3}Y-I_z+\frac{1}{2}) \qquad (6.5)$$
$$+X_{sb}(2I+\frac{2}{3}Y+1)(2I-\frac{4}{3}Y+1)$$

$$H'=K_{sbb}(2I-\frac{4}{3}Y)^{\frac{1}{2}}(J_+ + J_- + K_+ + K_-)+K_{st}(I_+ + I_-)$$

For this system, I_z and Y are not constants of motion. However, I_z is *approximately* conserved for Fermi resonance. This can be seen by directly calculating the expected values of I_z for $\Psi_1=(a_s^+ + a_t^+)\left|n_s,n_t,n_b\right\rangle$ and $\Psi_2=\left|n_s,n_t,n_b\right\rangle$:

$$\left\langle I_z\right\rangle_{\Psi_1}=(n_s-n_t)(n_1+n_2+3)/(n_1+n_2+2)\approx(n_s-n_t)$$

$$\left\langle I_z\right\rangle_{\Psi_2}=(n_s-n_t)$$

i.e.,

$$\left\langle I_z\right\rangle_{\Psi_1}\approx\left\langle I_z\right\rangle_{\Psi_2}$$

Hence, for strong Fermi resonance $(|K_{sbb}|>>|K_{st}|)$, I_z is approximately conserved or I_z is very small.

By Heisenberg's correspondence, we have the classical expression for H' :

$$2K_{sbb}[(I+\frac{1}{3}Y+I_z+\frac{1}{2})^{\frac{1}{2}}\cos(\varphi+\phi)/2+(I+\frac{1}{3}Y-I_z+\frac{1}{2})^{\frac{1}{2}}\cos(\varphi-\phi)/2](I-\frac{2}{3}Y+\frac{1}{2})$$

$$+2K_{st}[(I+\frac{1}{3}Y+\frac{1}{2})^2-I_z^{\,2}]^{\frac{1}{2}}\cos\phi \qquad (6.6)$$

for which,

$$\phi=\phi_s-\phi_t,\quad \varphi=\phi_s+\phi_t-4\phi_b$$

are the corresponding angles to I_z and Y, respectively.

Here, we note that the period of I_z and Y is 4π, and not 2π. In other words, the physical situations are different for ϕ or φ being 0 and 2π. This can be seen from the following analysis:

Suppose $\varphi = 0$, for $\phi = 0,\ 2\pi$, we have, respectively, that

$$\phi_s - 2\phi_b = 0,\ \phi_t - 2\phi_b = 0$$

and

$$\phi_s - 2\phi_b = \pi,\ \phi_t - 2\phi_b = -\pi$$

This means that as $\phi = 0$, the phase angle between *s* (*t*) and *b* is 0 while as $\phi = 2\pi$, the angle for *b* is in between those angles of *s* and *t*.

Similarly, suppose $\phi = 0$, as $\varphi = 0,\ 2\pi$, we have:

$$\phi_s - 2\phi_b = 0,\ \phi_t - 2\phi_b = 0$$

and

$$\phi_s - 2\phi_b = \pi,\ \phi_t - 2\phi_b = \pi$$

Obviously, these two situations are different.

Suppose the coefficients in the second quantized Hamiltonian are known, then from (6.4) and by choosing appropriate basis functions, the corresponding Hamiltonian matrix can be constructed. By diagonalizing the matrix, the eigenenergies can be obtained. By equating eigenenery *E* to (6.5) + (6.6), we have the classical dynamical equation for the system. Since the equation is implicit and there is coupling between ϕ and φ coordinates, its analytical solutions are difficult to obtain. In the following, we will try to solve the dynamics under strong Fermi resonance.

6.4 Dynamics under strong Fermi resonance

Under strong Fermi resonance, $|K_{sbb}| >> |K_{st}|$ or $\dot{Y} >> \dot{I}_z$. As a first step, in (6.5) + (6.6), *Y* can be approximated by its average $\langle Y \rangle$, i.e.,

$$\langle Y \rangle = \sum |C_i|^2\, Y_i$$

in which, C_i is the eigencoefficient for the eigenstate as:

$$\sum C_i |n_{s_i}\rangle |n_{t_i}\rangle |n_{b_i}\rangle$$

Y_i is

$$\frac{1}{2}(n_{s_i} + n_{t_i} - n_{b_i})$$

Due to strong coupling, we suppose that $Y(\varphi)$ is localized with φ being specified by certain values. By substituting the designated $\langle Y \rangle$ and φ in (6.5)+(6.6) and equating it to *E*, then $I_z(\phi)$ can be obtained. By substituting some values of $I_z(\phi)$ in (6.5)+(6.6), $Y(\varphi)$ can be finally obtained. For consistency, we may assume those $Y(\varphi)$ not satisfying the condition of strong coupling in (6.5)+(6.6) to check if the thus obtained $I_z(\phi)$ is congruent to the original trajectory. The thus obtained $I_z(\phi)$ is not very accurate, but if the calculation is consistent and the trajectories of $I_z(\phi)$ and $Y(\varphi)$ are stable with consistent characters, then they show the characters of the rigorous solutions.

Numerically, the following data are adopted (Refs.6.1,6.7–6.9): $\omega_s = \omega_t = 3028$ cm^{-1}, $\omega_b = 1391$ cm^{-1}, $X_{ss} = -28.45$ cm^{-1}, $X_{bb} = -0.16$ cm^{-1}, $X_{st} = 0.3$ cm^{-1}, $X_{sb} = -14.0$ cm^{-1}. $K_{st} = 5$ cm^{-1} and $K_{sbb} = -100$ cm^{-1} are taken to show strong Fermi resonance. Suppose $n_s + n_t + \frac{n_b}{2} = 7$ and that $Y(\varphi)$ is limited to $\varphi = 0$, π, 2π and 3π. In fact, solutions corresponding to $\varphi = 0$, π are independent and solutions to $\varphi = 2\pi$, 3π can thus be obtained. In all the 36 levels, there are 3 levels that can provide these solutions. Here, only one of them will be demonstrated. Fig.6.7 shows the $I_z(\phi)$ trajectory with $E = 25985.474$ cm^{-1}, $\langle Y \rangle = -0.3709$, $\varphi = 0$. Fig.6.8 shows the trajectories $Y(\varphi)$ by the eight points of $I_z(\phi)$ shown in Fig.6.7. Fig.6.9 shows $I_z(\phi)$ corresponding to the four points

$(Y, \varphi = 0)$ shown in Fig.6.8.

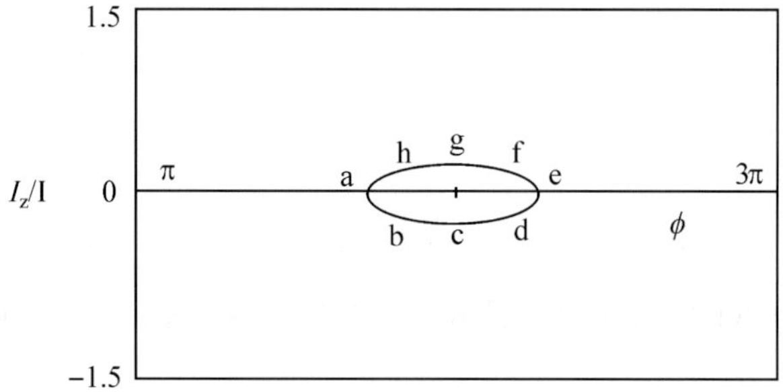

Fig.6.7 The $I_z(\phi)$ trajectory with $E = 25985.474$ cm^{-1}, $\langle Y \rangle = -0.3709$, $\varphi = 0$

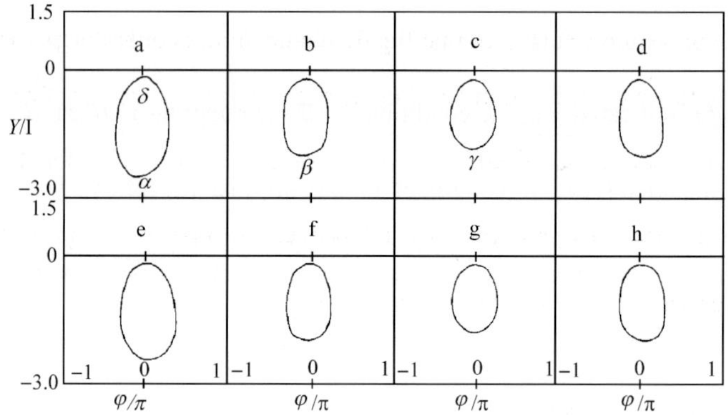

Fig.6.8 The trajectories thus obtained corresponding to the eight points of (I_z, ϕ) shown in Fig.6.7. See text for details.

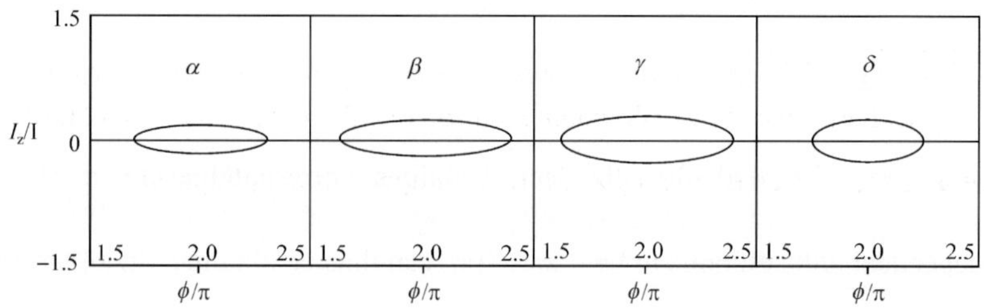

Fig.6.9 The $I_z(\phi)$ trajectories corresponding to the four points $(Y, \varphi = 0)$ shown in Fig.6.8. See text for details.

In Fig.6.8, as $\varphi = 0$ and Y is very small, the obtained $I_z(\phi)$ is still consistent with $I_z(\phi)$ as shown in Fig.6.7 though the strong condition $\dot{Y} >> I_z$ is not necessarily obeyed. Calculational consistency shows that $Y(\varphi)$ and $I_z(\phi)$ are localized around $\varphi = 0$, $\phi = 2\pi$ (or equivalently, $\varphi = 2\pi$, $\phi = 0$). Unexpectedly, there are no solutions for $\varphi = \pi$ (or 3π). From the Figures, it is seen that $\dot{Y} >> I_z$ and I_z is close to a constant during the dynamical evolution. The analysis shows that $Y(\varphi)$ and $I_z(\phi)$ are two localized, independent trajectories. For other states, $Y(\varphi)$ and $I_z(\phi)$ are intertwined, coupled and complicated trajectories.

The above analysis shows that the dynamics for a coupled three-mode system is very complicated. It is a three dimensional trajectory in the four dimensional (Y, φ, I_z, ϕ) space constrained by energy constancy. However, under strong coupling, there are states which possess two independent subspaces, $Y(\varphi)$ and $I_z(\phi)$. The trajectories are then reduced to two dimensions.

6.5 Semiclassical fixed point structure

The global dynamical structure for a Hamiltonian system is expressed in the distribution and properties of its fixed points. The fixed point structure for an su(3) system is what we want to explore. It will offer dynamical information for a three-mode system.

The harmonic and anharmonic terms for three identical oscillators in su(3) operators have been expressed as in (6.2). Suppose its couplings are:

$$t(a_1^+ a_2 + a_1^+ a_3 + a_2^+ a_3 + h.c.)$$

$$+\delta(a_1^+ a_1^+ a_2 a_2 + a_1^+ a_1^+ a_3 a_3 + a_2^+ a_2^+ a_3 a_3 + h.c.)$$

By employing Heisenberg's correspondence, the t term can be cast as:

$$2[(I+\frac{1}{3})Y^2 - I_z^2]^{\frac{1}{2}} \cos\phi_{12} + 2[(I+\frac{Y}{3}+I_z)(I-\frac{2}{3}Y)]^{\frac{1}{2}} \cos\phi_{13}$$

$$+2[(I+\frac{Y}{3}-I_z)(I-\frac{2}{3}Y)]^{\frac{1}{2}}\cos\phi_{23}$$

The δ term is

$$2[(I+\frac{Y}{3})^2-I_z{}^2]\cos(2\phi_{12})+2(I+\frac{Y}{3}+I_z)(I-\frac{2}{3}Y)\cos(2\phi_{13})$$
$$+2(I+\frac{Y}{3}-I_z)(I-\frac{2}{3}Y)\cos(2\phi_{23})$$

here, $\phi_{ij}=\phi_i-\phi_j$ .The corresponding phase angles for I_z , Y are: $\phi=\phi_{12}$, $\varphi=\phi_{13}+\phi_{23}$ (or $\phi_{12}=\phi$, $\phi_{13}=(\varphi+\phi)/2$, $\phi_{23}=(\varphi-\phi)/2$)

The Hamilton's equations are:

$$\partial H/\partial\phi=-\dot{I}_z\,,\partial H/\partial\varphi=-\dot{Y}\,,\partial H/\partial I_z=\dot{\phi}\,,\partial H/\partial Y=\dot{\varphi}$$

This will lead to:

$$\dot{I}_z=t\{2[(I+\frac{1}{3}Y)^2-I_z{}^2]^{\frac{1}{2}}\sin\phi+[(I+\frac{1}{3}Y+I_z)(I-\frac{2}{3}Y)]^{\frac{1}{2}}\sin(\varphi+\phi)/2$$
$$-[(I+\frac{1}{3}Y-I_z)(I-\frac{2}{3}Y)]^{\frac{1}{2}}\sin(\varphi-\phi)/2\}$$
$$+2\delta\{2[(I+\frac{1}{3}Y)^2-I_z{}^2]\sin 2\phi+(I+\frac{1}{3}Y+I_z)(I-\frac{2}{3}Y)\sin(\varphi+\phi)$$
$$-(I+\frac{1}{3}Y-I_z)(I-\frac{2}{3}Y)\sin(\varphi-\phi)\}$$

$$\dot{Y}=t\{[(I+\frac{1}{3}Y+I_z)(I-\frac{2}{3}Y)]^{\frac{1}{2}}\sin(\varphi+\phi)/2$$
$$+[(I+\frac{1}{3}Y-I_z)(I-\frac{2}{3}Y)]^{\frac{1}{2}}\sin(\varphi-\phi)/2\}$$
$$+\delta\{2(I+\frac{1}{3}Y+I_z)(I-\frac{2}{3}Y)\sin(\varphi+\phi)+2(I+\frac{1}{3}Y-I_z)(I-\frac{2}{3}Y)\sin(\varphi-\phi)\}$$

$$\dot{\phi}=2(2\alpha-\alpha')I_z+t\{(I-\frac{2}{3}Y)[(I+\frac{1}{3}Y+I_z)(I-\frac{2}{3}Y)]^{-\frac{1}{2}}\cos(\varphi+\phi)/2$$
$$-(I-\frac{2}{3}Y)[(I+\frac{1}{3}Y-I_z)(I-\frac{2}{3}Y)]^{-\frac{1}{2}}\cos(\varphi-\phi)/2\}$$
$$+2\delta\{(I-\frac{2}{3}Y)\cos(\varphi+\phi)-(I-\frac{2}{3}Y)\cos(\varphi-\phi)\}$$

$$\varphi = 2(2\alpha - \alpha')\frac{1}{3}Y + t\{\frac{2}{3}[(I + \frac{1}{3}Y)^2 - I_z{}^2]^{-\frac{1}{2}}(I + \frac{1}{3}Y)\cos\phi$$

$$-[(I + \frac{1}{3}Y + I_z)(I - \frac{2}{3}Y)]^{-\frac{1}{2}}(\frac{1}{3}I + \frac{4}{9}Y + \frac{2}{3}I_z)\cos(\varphi + \phi)/2$$

$$-[(I + \frac{1}{3}Y - I_z)(I - \frac{2}{3}Y)]^{-\frac{1}{2}}(\frac{1}{3}I + \frac{4}{9}Y - \frac{2}{3}I_z)\cos(\varphi - \phi)/2\}$$

$$+\delta\{[\frac{4}{3}(I + \frac{1}{3}Y)]\cos 2\phi - 2(\frac{1}{3}I + \frac{4}{9}Y + \frac{2}{3}I_z)\cos(\varphi + \phi)$$

$$-2(\frac{1}{3}I + \frac{4}{9}Y - \frac{2}{3}I_z)\cos(\varphi + \phi)\}$$

The fixed point structure is determined by $\dot{I}_z = 0$, $\dot{Y} = 0$, $\dot{\phi} = 0$, $\dot{\varphi} = 0$. It is a two dimensional set in four dimensional space (due to energy constancy). Since I_z, Y, ϕ, φ are intertwined, analytical solutions are impossible. This is different from the two-mode system with su(2) algebra (Ref.6.2) which is analytically soluble. Here, we will try the solutions with the constraint $I_z = 0$, which are the trajectories in two dimensional space.

From the Hamilton's equations and the constraint $I_z = 0$, the conditions for the fixed points are:

$$\phi = 0, \quad \frac{\varphi}{2} = 0, \quad \pi$$

and

$$\varphi = \alpha_0 Y + t\{\frac{2}{3} \mp [\frac{2}{3}(I + \frac{4}{3}Y)][(I + \frac{1}{3}Y)(I - \frac{2}{3}Y)]^{-\frac{1}{2}}\}$$

$$-\delta(\frac{4}{3}Y) = 0$$

Here, $\alpha_0 = 2(2\alpha - \alpha')/3$. $+$ and $-$ are for $\frac{\varphi}{2} = 0$ and π, respectively.

Furthermore, if $\delta = 0$, above equation is of the third order and is analytically soluble. For convenience, we define $\beta = t/(2\alpha - \alpha')$.

Fig.6.10 shows the fixed point structure for $I = 20$, $-30 < \beta < 30$ in which (a)

$\frac{\varphi}{2}=0$, (b) $\frac{\varphi}{2}=\pi$. For other *I*, the characters are similar to the case with I = 20 except explicit numerical values.

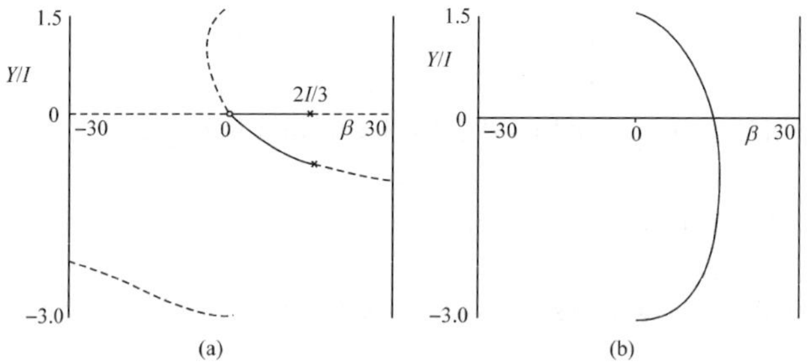

Fig.6.10 The fixed point structure for I = 20. (a) $\frac{\varphi}{2}=0$, (b) $\frac{\varphi}{2}=\pi$. * is the bifurcation point, o is the singular point, solid lines are the set of stable fixed points and the broken lines are of unstable fixed points.

In Fig.6.10, we note that:

(a) For $\frac{\varphi}{2}=0$, despite of *t*, δ, Y = 0 is always a solution. The bifurcation points satisfy the equations:

$$\varphi=0,\ \partial\varphi/\partial Y=0$$

Hence, for $Y = 0$, $\beta=\frac{2}{3}I$ is a bifurcation point. As $\beta>\frac{2}{3}I$, the fixed points are unstable (broken lines), i.e., $\partial\varphi/\partial Y>0$. As $0<\beta<\frac{2}{3}I$, they are stable, i.e., $\partial\varphi/\partial Y<0$. As $\beta<0$, fixed points are unstable. $\beta=0$ is not a bifurcation point. In fact, it is a singular point, i.e., its $\partial\varphi/\partial Y$ is divergent. Across the singular point, the stability changes.

(b) Consider $\frac{\varphi}{2}=0$ with $Y\neq 0$. For any *I*, as $\beta>0$, there is always a fixed point. As $\beta<0$, there are always one or three fixed points. For $\beta>0$, there are bifurcation points. Fig.(a) shows the case of I = 20, bifurcation appears at $\beta\approx 15.6$, $Y/I\approx -0.765$. As $0<\beta\leq 15.6$, fixed points are stable, while for $\beta<0$ and

$\beta \geq 15.6$ they are unstable.

(c) For $\frac{\varphi}{2} = \pi$, fixed points appear only in a certain range with $\beta > 0$. For I = 20, the range is between 0 and ≈ 14.4. Also, there are no bifurcation points.

(d) For $\frac{\varphi}{2} = 0$, $|\beta| >> 1$, $Y \approx -1.5I$ (and, of course, $Y = 0$), these fixed points correspond to the normal mode. As $\beta \approx 0$ and is negative, $Y = 0$, $1.5I$, $-3I$, these fixed points correspond to local mode. The condition with $\phi = 0$ and $\frac{\varphi}{2} = 0$ shows that the three oscillators are in phase. (That with $\phi = 0$ and $\frac{\varphi}{2} = \pi$ shows that the two oscillators are in phase while the third one is out of phase.) Suppose I = 4, the actions for the normal mode and three local modes are, respectively, (n_1, n_2, n_3) = (2, 2, 8) and (4, 4, 4), (0, 0, 12), (6, 6, 0).

Fixed point structure shows the global dynamical properties. That is, it demonstrates the dynamical properties as the coefficients of the system Hamiltonian vary. It offers us an overview of the system dynamics. This is its merit. For a three-mode system with su(3) algebra, simple analytical solutions are impossible though under some specific conditions, they may be possible. All these demand that we adhere to other approaches to explore the su(3) system.

References

6. 1 M. E. Kellman, *J. Chem. Phys*., 83(1985)3843.

6. 2 L. Xiao and M. E. Kellman, *J. Chem.Phys*., 90(1989)6086.

6. 3 M. E. Kellman and E. D. Lynch, *J. Chem. Phys*., 85(1986)7216.

6. 4 M. E. Kellman and E. D. Lynch, *J. Chem. Phys*., 88(1988)2205.

6. 5 Z. Li, L. Xiao and M. E. Kellman, *J. Chem. Phys*., 92(1990)2251.

6. 6 V. A. Dulock and H. V. McIntosh, *Am. J. Phys*., 33(1965)109.

6. 7 M. E. Kellman, *J. Chem. Phys*., 93(1990)6630.

6. 8 O. Mortensen, B. R. Henry and M. A. Mohammadi, *J. Chem. Phys*.,75 (1981) 4800

6. 9 Z. Chila and A. Chedin, *J. Mol. Spectrosc*., 40(1971)337.

6. 10 G. Wu, *Chem. Phys*., 173(1993)1.

6. 11 G. Wu, *Chem. Phys. Lett*., 195(1992)115.

6. 12 G. Wu, *Chem. Phys. Lett*., 179(1991)29.

Chapter 7

Application of su(3) algebra

7.1 su(3) algebraic method

The dynamics of coupled three-mode oscillators is more complex than a two-mode system. The main reason is that it can be nonintegrable and chaotic. Its coset space is four dimensional, in which different initial points will lead to different trajectories according to the Hamilton's equations of motion. Before discussing the real systems of three-mode oscillators like H_2O, we will first introduce the su(3) algebraic methods with simulation.

Consider the Hamiltonian of three-mode oscillators for which two modes are identical:

$$
\begin{aligned}
&\omega_s(n_s+n_t+1)+\omega_b(n_b+\frac{1}{2})\\
&+X_{ss}[(n_s+\frac{1}{2})^2+(n_t+\frac{1}{2})^2]+X_{bb}(n_b+\frac{1}{2})^2\\
&+X_{st}(n_s+\frac{1}{2})(n_t+\frac{1}{2})+X_{sb}(n_s+n_t+1)(n_b+\frac{1}{2})\\
&+K(a_s^+a_b+a_t^+a_b+a_s^+a_t+h.c.)
\end{aligned}
$$

Here, ω_s, ω_b, X_{ss}, X_{bb}, X_{st}, X_{sb} are the linear and nonlinear coefficients. K is the coupling coefficient.

We may cast this Hamiltonian in the coordinates (q_t, p_t, q_b, p_b) by the coset representatives. That is:

$$
n_s=N-\frac{1}{2}(q_t^2+p_t^2+q_b^2+p_b^2)
$$

$$
n_t=(q_t^2+p_t^2)/2 \qquad n_b=(q_b^2+p_b^2)/2
$$

with $N=n_s+n_t+n_b$. As an approximation, n_α^2, $n_\alpha n_\beta$ can be taken as the products of the corresponding n_α, n_β. The coupling term is:

$$K[(q_t+q_b)[2N-\sum_{\alpha=t,b}(q_\alpha^2+p_\alpha^2)]^{\frac{1}{2}}+(q_tq_b+p_tp_b)]$$

The Hamilton's equations of motion are ($\partial H/\partial p_\alpha = dq_\alpha/dt, \partial H/\partial q_\alpha = -dp_\alpha/dt$):

$$\dot{q}_t = (-2X_{ss}+X_{st})[N-(q_t^2+p_t^2)-\frac{1}{2}(q_b^2+p_b^2)]p_t$$
$$+K\{-[2N-(q_t^2+p_t^2)-(q_b^2+p_b^2)]^{-1/2}(q_t+q_b)p_t+p_b\}$$

$$\dot{p}_t = (2X_{ss}-X_{st})[N-(q_t^2+p_t^2)-\frac{1}{2}(q_b^2+p_b^2)]q_t$$
$$+K\{-[2N-(q_t^2+p_t^2)-(q_b^2+p_b^2)]^{1/2}$$
$$+[2N-(q_t^2+p_t^2)-(q_b^2+p_b^2)]^{-1/2}(q_t^2+q_tq_b)-q_b\}$$

$$\dot{q}_b = (\omega_b-\omega_s)p_b+X_{ss}[-2N+(q_t^2+p_t^2)+(q_b^2+p_b^2)-1]p_b$$
$$+X_{bb}[(q_b^2+p_b^2)+1]p_b-\frac{1}{2}X_{st}[(q_t^2+p_t^2)+1]p_b$$
$$+X_{sb}[N+\frac{1}{2}-(q_b^2+p_b^2)]p_b$$
$$+K\{p_t-p_b(q_t+q_b)[2N-(q_t^2+p_t^2)-(q_b^2+p_b^2)]^{-1/2}\}$$

$$\dot{p}_b = (\omega_s-\omega_b)q_b+X_{ss}[2N-(q_t^2+p_t^2)-(q_b^2+p_b^2)+1]q_b$$
$$-X_{bb}[q_b^2+p_b^2+1]q_b+\frac{1}{2}X_{st}[q_t^2+p_t^2+1]q_b$$
$$-X_{sb}[N+\frac{1}{2}-(q_b^2+p_b^2)]q_b$$
$$-K\{[2N-(q_t^2+p_t^2)-(q_b^2+p_b^2)]^{1/2}+(q_t+q_b)(-q_b)[2N-(q_t^2+p_t^2)-(q_b^2+p_b^2)]^{-1/2}+q_t\}$$

Hence, as an initial $(q_t, p_t, q_b, p_b)_0$ is given, the following dynamical trajectory is known. As an example, we take $\omega_s=\omega_b=3028$ cm^{-1}, X_{ss}, X_{bb}, X_{st}, X_{sb} are all 0, K = -100 cm^{-1}, N = 3. The energy level is 13767.4 cm^{-1}. Initial values are: $q_t^0=p_b^0=0$, $p_t^0=1.4376$, $q_b^0=$ -0.7746 and $q_t^0=0$, $p_t^0=0.0411$, $q_b^0=$ -0.7746, $p_b^0=1.3416$. The actions of the trajectories as a function of time are shown in Fig. 7.1(a), (b). Obviously, the actions and phases of these trajectories are quite different. The phase angles between s and t (ϕ_{st} (tan^{-1}-p_t/q_t)) cover the whole range of (0, 2π), while the angles between s and b (ϕ_{sb} ($\tan^{-1}$-p_b/q_b)) are localized. This is different from the case of an integrable two-mode system. For such an su(3) system, there is no way to interpret whether a level is local or normal.

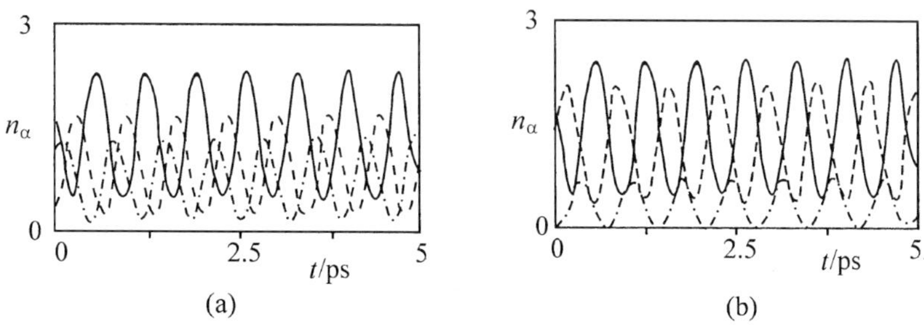

Fig.7.1 The evolution of n_s (———), n_t (---------), n_b (·········). See text for details.

As a second example, take ω_s= 3028 cm^{-1}, ω_b= 3477.5 cm^{-1}, X_{ss}, X_{bb}, X_{st} and X_{sb} as 0. K = −100 cm^{-1} with coupling only between s, b and t, b. Initial actions are $n_s^0 = 10$, $n_t^0 = n_b^0 = 0$. Fig.7.2 shows the evolution of n_s, n_t, n_b which indicates that action transfer is only significant between s and t. The b oscillator acts as a pump of period 0.396 ps and does not gain significant action. As ω_s= 3028 cm^{-1}, ω_b= 1391 cm^{-1}, the period drops to 0.126 ps. In this case, the resonance between b and s, t is no longer obvious. The phenomenon that b acts as a pump has been reported by Kommandeur and Zewail (Ref.7.1). The associated peak is 200 cm^{-1}, corresponding to a period of 0.2 ps. This value is very close to what we simulated here. We note also that the phase angle range between s and b is (0, 2π) while that between s and t is locked at $\pi/2$ and $3\pi/2$.

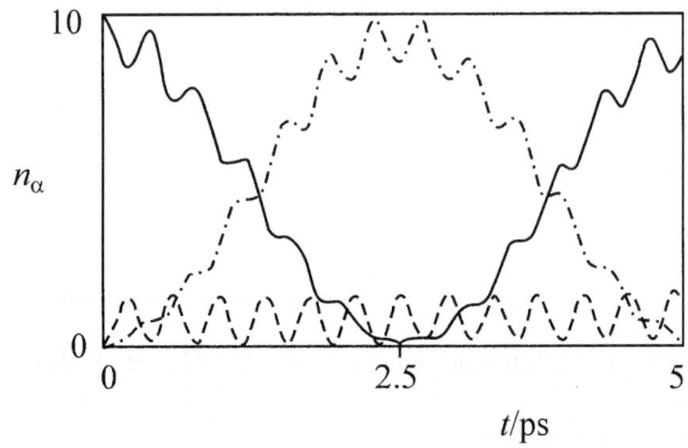

Fig.7.2 The evolution of n_s (———), n_t (- - - - -), n_b (- - - - -). See text for details.

Fig.7.3 shows another case. Parameters are $n_s^0 = n_t^0 = 4$ ($q_t^0 = 0$, $p_t^0 = \sqrt{8}$), $n_b^0 = 0$, $\omega_s = \omega_t = 3028$ cm^{-1}, $\omega_b = 2782$ cm^{-1}, $K = -100$ cm^{-1}. At the beginning, the actions on the *s* and *t* oscillators are the same. Subsequently, the action is transferred *to and fro* between the two oscillators. If initially, $q_t^0 = \sqrt{8}$, $p_t^0 = 0$, the actions on *s* and *t* simultaneously flow *to and out of* the *b* oscillator. With regard to the angle, ϕ_{sb} is not localized while ϕ_{st} is localized around $\pi/2$ and $3\pi/2$.

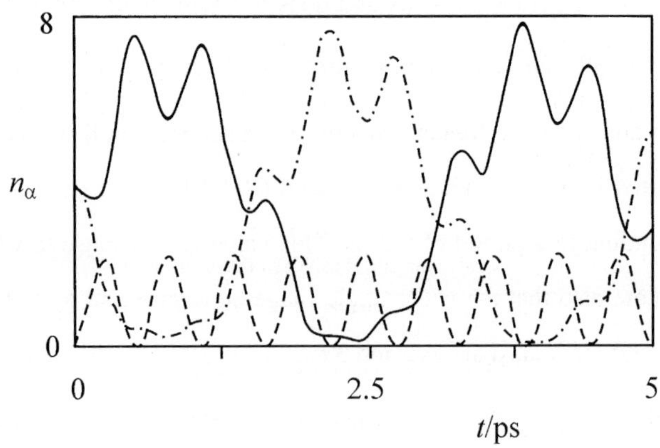

Fig.7. 3 The evolution of n_s (———), n_t (- - - - -), n_b (- - - - -). See text for details.

7.2 Fitting of the coefficients

For the three-atomic molecules, such as H_2O, D_2O, H_2S, CH_2- of CH_2Br_2 and CD_2- of CD_2Br_2 (Since Br is much heavier than C, H, D, CH_2- and CD_2- can be considered as single entities), there are two identical bond stretches and one bend. The couplings are 1:1, 2:2 (Darling-Dennison) and Fermi resonance between the stretches and the bend. The algebraic Hamiltonian is $H = H_0 + H_r$. H_0 is the anharmonic term and H_r the coupling term:

$$H_0 = \omega_s(n_s + n_t + 1) + \omega_b(n_b + \frac{1}{2}) + X_{ss}[(n_s + \frac{1}{2})^2 + (n_t + \frac{1}{2})^2]$$
$$+ X_{bb}(n_b + \frac{1}{2})^2 + X_{st}(n_s + \frac{1}{2})(n_t + \frac{1}{2}) + X_{sb}(n_s + n_t + 1)(n_b + \frac{1}{2})$$
$$H_r = K_{st}(a_s^+ a_t + h.c.) + K_{DD}(a_s^+ a_s^+ a_t a_t + h.c)$$
$$+ K_F(a_s^+ a_b a_b + a_t^+ a_b a_b + h.c.)$$

For these couplings, $N = n_s + n_t + n_b/2$ is a constant.

As shown in Section 1.4, these coefficients can be determined by the fitting of the eigenvalues of the algebraic Hamiltonian to the experimental data. Experimental data for H_2O, D_2O, H_2S are from Ref.7.2; for CH_2Br_2 from Ref. 7.3 and for CD_2Br_2 from Refs.7.4 and 7.5. The fitted coefficients are listed in Table7.1. The calculated and experimental level energies are listed in Appendices. The averaged deviations are, respectively, 7.2 cm^{-1} (H_2O), 6.3 cm^{-1} (D_2O), 3.0 cm^{-1} (H_2S), 3.7 cm^{-1} (CH_2Br_2) and 2.4 cm^{-1} (CD_2Br_2).

Table7.1 The coefficients of the algebraic Hamiltonian. The unit is cm^{-1}

	H_2O	D_2O	CH_2Br_2	CD_2Br_2	H_2S
ω_s	3890.64	2836.67	3174.90	2337.02	2735.15
ω_b	1645.25	1204.61	1439.89	1072.76	1229.83
X_{ss}	-82.06	-44.15	-68.71	-36.81	-48.69
X_{bb}	-16.18	-7.02	-6.85	-7.43	-9.57
X_{st}	-13.27	-10.57	-0.05	-4.84	-4.06

X_{sb}	-21.03	-13.14	-26.78	-16.04	-22.76
K_{st}	-42.80	-54.65	-39.32	-60.65	-8.11
K_{DD}	-0.19	-0.02	-4.00	-2.36	-0.58
K_F	-14.58	-5.83	-21.61	-17.93	-23.16

From these coefficients, we may conclude that:

(1) X_{ss}, X_{bb} are negative. This shows the character of a Morse oscillator.

(2) $|X_{ss}|>|X_{sb}|$, $|X_{bb}|\geq|X_{st}|$

(3) $|K_{st}|$ is only smaller than $|X_{ss}|$ and, in certain cases, even larger than $|X_{ss}|$ (except H_2S). For H_2S, Fermi resonance is evident.

(4) 2: 2 resonance is not eminent and is negligible while Fermi resonance is not negligible.

(5) By the comparison of CH_2- and CD_2-, it can be concluded that $|X_{ss}|$ decreases by deuteration. This is because that deuteration reduces the stretch amplitude, hence the anharmonicity $|X_{ss}|$. In addition, we note that deuteration does not affect Fermi resonance much, instead, it enhances the 1:1 resonance. These tendencies are also true for H_2O and D_2O.

7.3 Dynamical properties

For three-mode oscillators like H_2O, there are two forms of the algebraic Hamiltonian by the coset representation:

(1) $n_s = N - n_t - n_b/2$, $n_t = (q_t^2 + p_t^2)/2$, $n_b = (q_b^2 + p_b^2)/2$

The coupling terms are:

$$K_{st}(2n_s)^{\frac{1}{2}}q_t$$

$$K_{DD}n_s(q_t^2 - p_t^2)$$

$$K_F\{\sqrt{n_s}(q_b^2 - p_b^2) + [q_t(q_b^2 - p_b^2) + 2p_tq_bp_b]/\sqrt{2}\}$$

(2) $n_b = 2(N - n_s - n_t)$， $n_s = (q_s^2 + p_s^2)/2$， $n_t = (q_t^2 + p_t^2)/2$

The couplings terms are:

$$K_{st}(q_s q_t + p_s p_t)$$

$$K_{DD}[\frac{1}{2}(q_s^2 - p_s^2)(q_t^2 - p_t^2) + 2q_s q_t p_s p_t]$$

$$K_F\sqrt{2n_b}(q_s + q_t)$$

These two expressions are equivalent. They can be chosen for convenience. (See Note.)

By these expressions with given coefficients and initial conditions, the dynamical trajectories can be obtained via the Hamilton's equations of motion,. This system dynamics is complicated, its trajectories are initial-condition dependent. This is also related to the fact that by the given n_s^0, n_t^0, n_b^0, there are multiple (q_α^0, p_α^0) and hence, multiple trajectories.

For simplicity, we will first discuss the case of action transfer from t to b oscillators for which $n_t^0 = n_b^0 = 0$ with various n_s^0. By the first expression, the corresponding $q_t^0, p_t^0, q_b^0, p_b^0$ are all zero. In the sub-picosecond time scale, n_s, n_t, n_b show motions close to periodic behavior. Their amplitudes reflect the transfer strength among the oscillators. In fact, the resonance strength is dependent on n_s^0. Fig.7.4 shows the ratios of actions transferred by these two resonances (i.e., $\Delta n_t / N$， $\Delta n_b / N$) for various n_s^0. For H_2O, as $n_s^0 \leq 2$, the resonance between s, t oscillators dominates while as n_s^0>2 Fermi resonance dominates. The case of CH_2- is similar to H_2O for which Fermi resonance is only stronger than that between s and t oscillators as n_s^0>2. For CD_2-, the resonance between s, t oscillators is always stronger than Fermi resonance. As n_s^0 is close to 10, these two resonances are of comparative strengths for all the systems in the sub-picosecond time scale . In a longer time scale, this conclusion may vary.

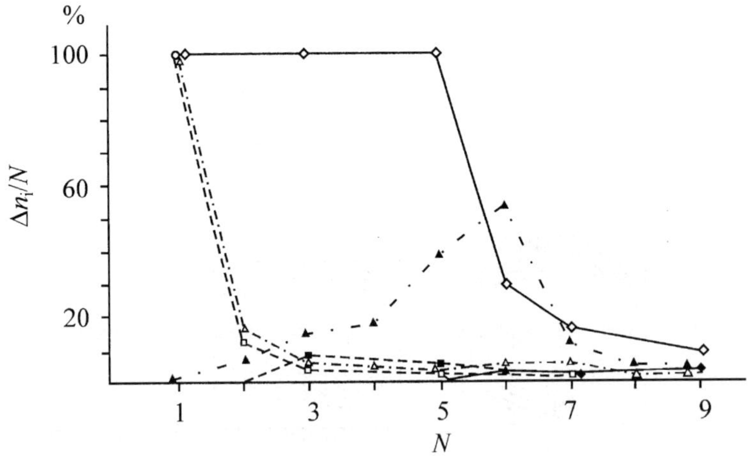

Fig.7.4 The variation of $\Delta n_t / N$ (open notation), $\Delta n_b / N$ (solid notation) as $n_s^0 = N$, for H_2O (△, ▲), CH_2- (□, ■), CD_2- (◇, ◆) See text for details.

Another point that needs attention is that although K_{DD} is small, as N is large 2:2 resonance can be significant and not negligible. This is evident as 2:2 resonance is proportional to q_α^4 (or p_α^4) while 1:1 resonance or Fermi resonance is proportional to q_α^3 (p_α^3).

7.4 Coset potential

In the Hamiltonian $H(q_t, p_t, q_b, p_b)$ or $H(q_s, p_s, q_t, p_t)$, let p_α be zero, then from $E = H(q_t, q_b)$ or $E = H(q_s, q_t)$, we can obtain $q_t = q_t(q_b)$ or $q_t = q_t(q_s)$. Here, E is the eigenenergy. Fig.7.5 shows the $q_t(q_s)$, $q_t(q_b)$ curves (The numbers show the levels counted from the lowest one) as $N = 4$ for H_2O, CH_2-, CD_2-. The potential formed by these curves can be called the coset potential.

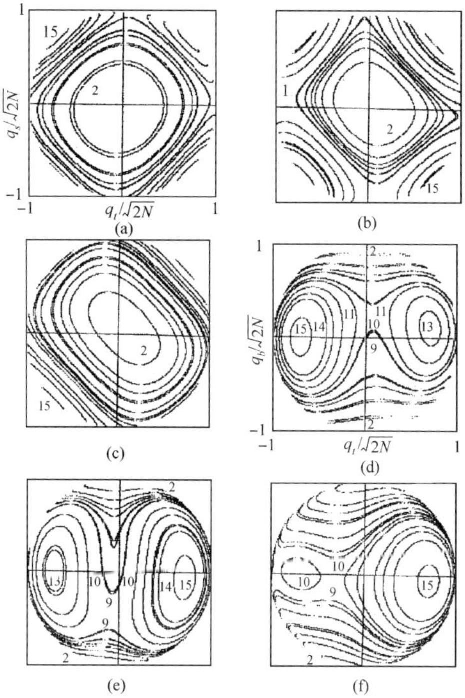

Fig.7.5 The potential curves for H_2O (a)(d), CH_2- (b)(e) and CD_2- (c) (f)

In Fig 7.5.(a)–(c), the symmetry by interchanging q_s, q_t is obvious. In Fig. 7.5 (d)–(f), the more asymmetric with respect to $q_b \leftrightarrow -q_b$, the more prominent is the Fermi resonance. As the pattern in Fig.7.5 is more symmetric under $q_b \leftrightarrow -q_b$ and more asymmetric under $q_t \leftrightarrow -q_t$, the more prominent is the resonance between *s* and *t* oscillators. This can be seen from the first kind of coset representation (See the last section). Hence, for the $9^{th}-13^{th}$ levels of H_2O, Fermi resonance is prominent while for the 14th and 15th levels, *s* and *t* resonance is prominent. The case of CH_2- is similar. For CD_2-, deuteration inhibits Fermi resonance and enhances the resonance between the *s* and *t* oscillators, especially for high levels. In summary, low levels do not show which resonance is leading while for high levels, either the resonance between *s* and *t* oscillators or Fermi resonance will show its importance more prominently.

Fig.7.6(a),(b) show the potentials as a Morse oscillator for CD_2 and CH_2.

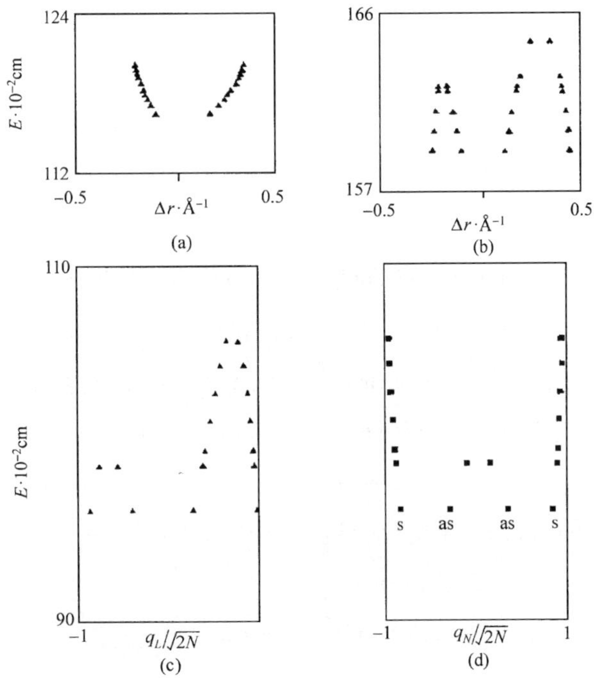

Fig.7.6 The potential curves. See text for details.

In the Figure, (a) is derived from Fig.7.5(c) with q_t=0. (b) is from Fig.7.5(e) with q_b=0. For the calculation of potential, see Section 1.2.

In Fig.7.6(b), there are two *wells* below the 13th levels while for levels above there is but one. A two-well pattern implies the local mode while a one-well pattern shows the normal mode (See Section 4.6). This can be understood from the analogy with the two-mode system. For the two-mode system, we have two dynamical situations: the local mode and normal mode. These two situations can be interchanged by the rotation of $\pi/2$ in the SU(2) coset space along the J_y axis (See Section 4.3). Fig. 7.6(c), (d) are the coset potentials for the local and normal mode pictures (coordinates are q_L, q_N) for O_3 with $N = 9$. In Fig. 7.6(c), those levels that support two wells are local modes while those that support one well with higher energies are normal. In the normal mode, the small well on the left side of Fig.7.6 (c) transforms to the central portion of antisymmetry in Fig. 7.6 (d) (labeled as) while the well in the right portion of Fig. 7.6 (c) transforms to the parts on the sides with larger q_N, showing symmetry (labeled s) in Fig. 7.6 (d). From Fig. 7.6 (d), we see that as levels are higher, the degree of antisymmetry diminishes until they become symmetric. By comparing Fig. 7.6 (b) with Fig. 7.6 (c), and Fig. 7.6 (d), we know that for the CH_2-system, as the levels are lower, symmetry and antisymmetry coexist and levels are local modes. For higher levels, only normal and symmetric modes survive. For H_2O, the 14th and 15th levels have $q_t < 0$, (Fig.7.5(d)) (while for CH_2-, the 14th, 15th levels have $q_t > 0$ (Fig.7.5(e))). Hence, the high levels of H_2O are antisymmetric. By comparing Fig.7.5(e),(f), it is known that for CD_2-, the 11th–15th levels are symmetric normal modes. By comparison, for CH_2- only the 14th and 15th levels are normal. Hence, deuteration enhances the vibrational normality.

We note that the vibrational normality, locality, symmetry and antisymmetry are only the dynamical properties with respect to the specific surfaces of section of S^3 (three dimensional sphere in four dimensional space) for the three mode vibrational system. Various surfaces may lead to different conclusions. Hence, they are the *local* properties in the phase space, not the *global* properties. Besides, we note that in su(3) algebra, there are no equivalent normal and local mode pictures. This is different from the two-mode system. It is an unsolved issue how to elucidate the dynamical properties on an arbitrary surface of section in the S^3 coset space. This *global* property is a fundamental issue for the vibrational dynamics that remains to be solved.

7.5 Statistical interpretation of locality and normality

In the su(3) coset space, different initial points will lead to different trajectories. Hence, for the coset phase space of a level, the phase angles of some initial points may be localized in certain ranges while some others may be extensive in $(0,2\pi)$. This is different from the two-mode system. In the integrable su(2) algebraic system every energy level is associated with a definite trajectory and definite phase angle range, either localized or extensive. In other words, the concept of local/normal modes is no longer applicable in an su(3) system. This concept is *local* in the phase space and not *global*. However, the following analysis will demonstrate that in a statistical way, for the levels of an su(3) system, the concept of local/normal modes is still valid in a global sense in the phase space.

The analysis first generates randomly q_t^0, q_b^0, p_b^0, then by $H(q_t^0, p_t^0, q_b^0, p_b^0) =$ eigenenergy to obtain p_t^0. Initial $(q_t^0, p_t^0, q_b^0, p_b^0)$ is used then to calculate the trajectory. The phase angle between s and t oscillators is $\tan^{-1}(-p_t / q_t)$ while that between s and b is $\tan^{-1}(-p_b / q_b)$. We will consider, in the time interval of 3.33 ps, the phase angle range. We consider the localized phase angle as normality. The calculation shows the percentage of normal trajectories by 100 randomly chosen initial points for an eigenstate. Calculation with more initial points, e.g. up to 500, offers the same result.

Calculation shows that ϕ_{sb} for H_2O, CH_2- and CD_2- are extensive in $(0,2\pi)$ for all the levels considered. For ϕ_{st}, the normality percentage increases from the low levels to high levels. In the statistical sense, this result is consistent with that in the su(2) system. Fig.7.7 shows the normality percentage (η) of the 15 levels with $N = 4$.

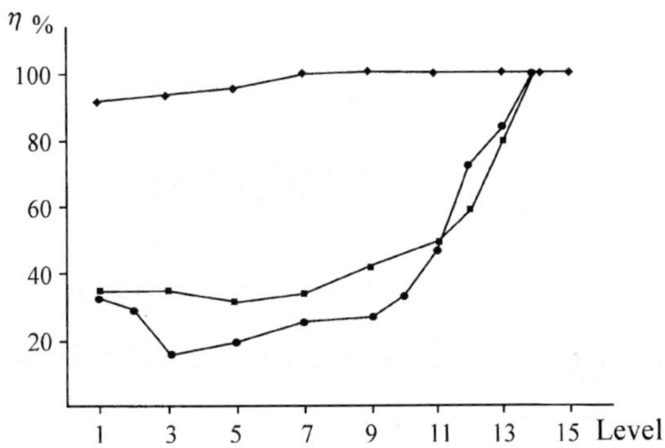

Fig.7.7 The normality percentage (η) of the 15 levels with $N = 4$: H_2O (■), CH_2- (●) and CD_2- (◆)

The normality of CD_2- is obviously larger than that of H_2O or CH_2-. As the level is above the 7th level, ϕ_{st} shows complete normality. Indeed, deuteration enhances normality. For H_2O and CH_2-, below the 11th level H_2O possesses larger normality while above the 11th level, the situation is opposite in that CH_2- possesses larger normality. Formally, we may consider a level with 50% or more normality as a normal mode while that with 50% or less normality as a local mode. In this way, we may say that for CD_2-, *s* and *t* oscillators always form a normal mode. For H_2O and CH_2-, levels below the 11th level are local while those above are normal.

In summary, the concept of local/normal modes in an su(2) system can be valid in the nonintegrable su(3) system, provided that it is interpreted in a global sense in the phase space.

7.6 Spontaneous symmetry breaking of identical modes

The two *s* and *t* stretches of H_2O-like molecules are identical. The Hamiltonian in (2) coset representation in Section 7.3 also shows symmetry under indices *s* and *t* interchanged. In general, an initial $(q_s^0, p_s^0, q_t^0, p_t^0)$ may not satisfy symmetry under *s* and *t* interchanged. However, after an interval of time, the trajectory always shows this symmetry

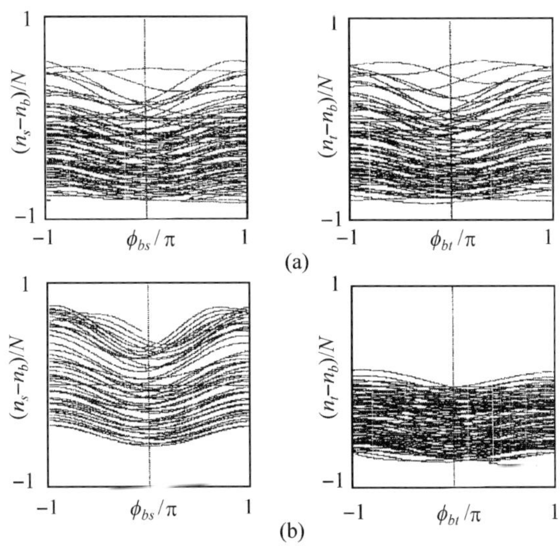

Fig.7.8 (a)Equivalent and (b) non-equivalent trajectories. See text for details.

Fig.7.8(a) shows the relations of $(n_s-n_b)/N$ versus ϕ_{bs} and $(n_t-n_b)/N$ versus ϕ_{bt} of the trajectory after 33.3 ps for H_2O, $N = 4$, energy 17688.23 cm^{-1}, initial conditions: $n_t^0 = 8.016\ 10^{-2}$, $\phi_{bt}^0 = 151.75^\circ$, $n_s^0 = 0.8439$, $\phi_{bs}^0 = -163.26^\circ$. Both are identical, as expected. However, there are the opposite cases as shown in Fig. 7.8(b) for which the initial conditions are: $n_t^0 = 4.0229\ 10^{-4}$, $\phi_{bt}^0 = 144.66^\circ$, $n_s^0 = 0.9142$, $\phi_{bs}^0 = 9.28^\circ$. In the Figure, n_t is apparently smaller than n_b while n_s and n_b are roughly the same in average.

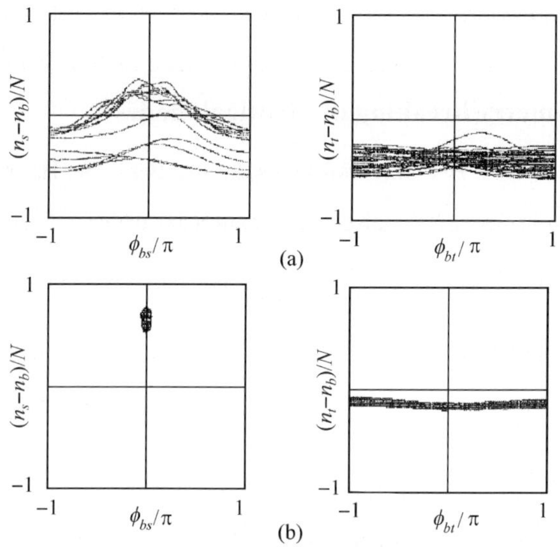

Fig.7.9 (a) The asymmetry trajectory and (b) localized excitation. See text for details

Fig.7.9 (a) shows the trajectory for CH_2-, $N = 4$, energy 15862.53 cm^{-1}, initial conditions: $n_t^0 = 0.2026$, $\phi_{bt}^0 = 31.74^\circ$, $n_s^0 = 1.8308$, $\phi_{bs}^0 = -103.25^\circ$ for which the asymmetry between s, t is so apparent. Fig.7.9(b) shows the trajectory for H_2O, $N = 4$, K_F is up to 100 cm^{-1}, energy 19447.45 cm^{-1}, initial conditions: $n_t^0 = 1.6089\ 10^{-2}$, $\phi_{bt}^0 = 128.69^\circ$, $n_s^0 = 3.3455$, $\phi_{bs}^0 = -6.34^\circ$. The increase of K_F causes ϕ_{bs} to be localized around 0. Also, n_s, n_t and n_b are fixed at certain values. This is a typical case of localized excitation for which the symmetry breaking between the *s* and t oscillators is very serious.

The above analysis shows that though the system Hamiltonian is symmetric under s and t interchange, its solutions or trajectories do not always possess this symmetry. This symmetry breaking originates from the intrinsic dynamical property, not from the external effects. Hence, it is spontaneous symmetry breaking.

7.7 Global symmetry and antisymmetry

Symmetric and antisymmetric vibrations are the main concepts in the traditional normal mode analysis. These concepts are no longer valid in the su(3) system. However, in the global sense in the coset space, they are still applicable according to a new interpretation as shown in this section.

For an eigenstate of the su(3) system with energy E, we can obtain the coset space solutions from $H(q_\alpha, p_\alpha) = E$. From (q_α, p_α), we can calculate (ϕ_{st}, ϕ_{sb}) and (ϕ_{bt}, ϕ_{bs}). These two relations will reveal the symmetry of the eigenstate. The results show that for H_2O, CH_2- and CD_2-, except at higher levels, the solution patterns of (ϕ_{st}, ϕ_{sb}) and (ϕ_{bt}, ϕ_{bs}) are composed of randomly distributed points. Fig.7.10 shows the case of the 14th level of H_2O, CH_2- and CD_2- with $N = 4$.

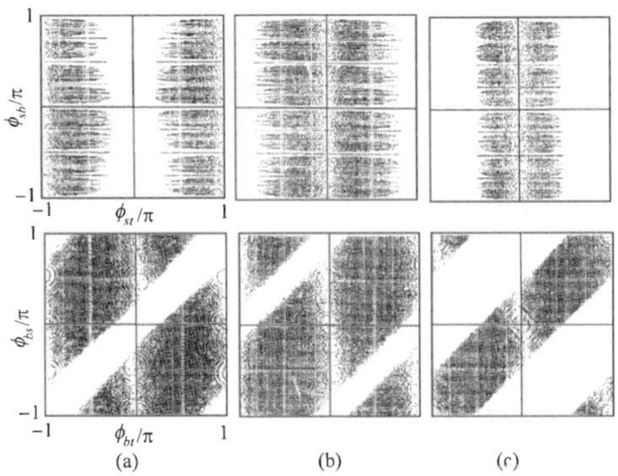

Fig.7.10 The solution patterns of the 14th level of H_2O(a), CH_2- (b) and CD_2- (c)

For CH_2-, CD_2-, in (ϕ_{st}, ϕ_{sb}), ϕ_{st} is centred around 0 while for H_2O, it is centred around π ($-\pi$). In (ϕ_{bt},ϕ_{bs}), for CH_2- and CD_2- the patterns are centred along $\phi_{bt}=\phi_{bs}$ while for H_2O it is along $\phi_{bt}-\phi_{bs}=\pi$. These observations show that for CH_2- and CD_2-, the two stretches are more or less in the symmetric phase relation while for H_2O they are in the antisymmetric phase relation. We note that different trajectories will have different phase relations. Hence, we cannot attribute to each state of an su(3) system a definite symmetric or antisymmetric mode. This is different from the su(2) system. However, from the global viewpoint in the phase space, though different trajectories will possess different phase angles, for CH_2- and CD_2- systems, there will be more possibility that the two stretches are in the symmetric phase relation while for H_2O, there will be more possibility that they are in the antisymmetric phase relation. Between the bend and stretch, there is no fixed phase angle. (The range of ϕ_{sb},ϕ_{bt} is from 0 to 2π. This is related to K_F being smaller than K_{st}) Hence, we say that the classical concept of symmetry and antisymmetry of the normal mode is still meaningful in the su(3) system from a global viewpoint in the phase space.

The above analysis is consistent with that shown in Section 7.4 by the coset potential.

7.8 Action transfer coefficient

The action transfer Δn_t and Δn_b among the s, t and b oscillators can be written as:

$$|\Delta n_t|=|q_t\dot{q}_t+p_t\dot{p}_t|\Delta t$$

$$=|q_t\partial H/\partial p_t-p_t\partial H/\partial q_t|\Delta t\equiv d_{st}\Delta t$$

and

$$|\Delta n_b|=|q_b\dot{q}_b+p_b\dot{p}_b|\Delta t$$

$$=|q_b\partial H/\partial p_b-p_b\partial H/\partial q_b|\Delta t\equiv d_{sb}\Delta t$$

Thus defined, d_{st} and d_{sb} show the action transfer coefficients among the oscillators. For a solution point (q_α, p_α) in the phase space, we have the corresponding d_{st} and d_{sb}. If they are averaged, then we have the global action transfer coefficients D_{st} and D_{sb}.

Fig.7.11 shows D_{st} and D_{sb} of the 15 levels for H_2O, CH_2- and CD_2-, with N = 4 (for a given N, there are $(N+1)(N+2)/2$ states)

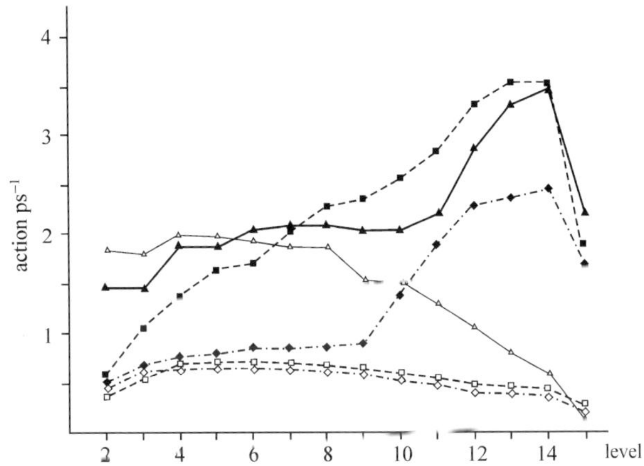

Fig.7.11 D_{st} (solid notations), D_{sb} (open notations) of H_2O (▲, △), CH_2- (◆, ◇), CD_2- (■, □)

From Fig. 7.11, we may deduce that (1) as a level is higher, D_{sb} is smaller while D_{st} is larger. This is related to the fact that for low states there are more actions in the *b* oscillator, while for high states there are more actions in the *s* and *t* oscillators. In general, $D_{st} > D_{sb}$. This shows that the action transfer between the *s* and *t* oscillators is more prominent than with the *b* oscillator. (2) From the low to high levels, the variation of D_{sb} of H_2O is more obvious than in the case of CH_2- and CD_2-. In the low levels of H_2O, D_{sb} is larger than D_{st}. This shows that for H_2O, Fermi resonance is important. (3) By comparing the cases of CH_2- and CD_2-, we know that deuteration does not affect D_{sb} much. Instead, it enhances D_{st} greatly.

7.9 Relaxational probability

With the concept of d_{st}, d_{sb} defined previously, we can define the relaxational probability of action transfer from *s* to *t* and *b* oscillators: they are P_{st}, P_{sb}. For a solution (q_t, p_t, q_b, p_b) in the coset space corresponding to an energy level, as d_{st} or d_{sb} is negative, then at time *t* (just Δt in the last section) if $|d_\alpha t/N| < 0.01$, we say that the state configuration survives and let $P_\alpha(t) = 1$. Otherwise, it does not survive and $P_\alpha(t) = 0$. After taking an average over all the solutions, we then have $P_{st}(t), P_{sb}(t)$. They show the survival probabilities at time *t* for the inter-oscillator relaxation.

Fig.7.12 shows $P_{st}(t)$ of the 11th level of CH_2- and $P_{sb}(t)$ of the 14th level of H_2O with $N = 4$.

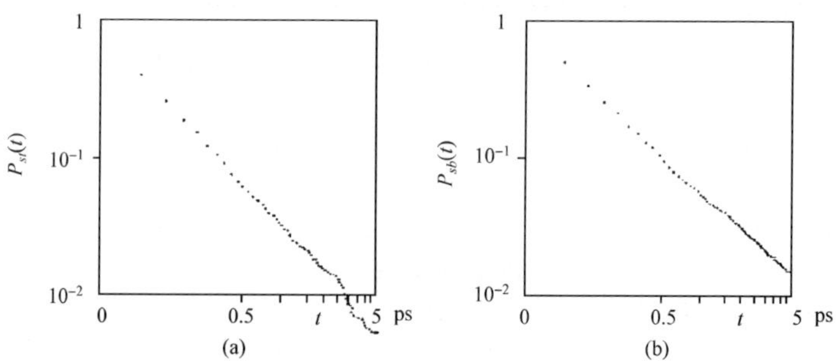

Fig.7.12 The survival probability for the levels of CH_2- (a), H_2O(b). See text for details.

From Fig. 7.12, we see that at $t = 0.5$ ps, there is already much relaxation. Both $P_{st}(t)$ and $P_{sb}(t)$ show power law behavior as t^{-d}. At $t < 0.5$ ps, *d* is close to 1. As *t* reaches 5 ps, *d* apparently deviates from 1. The *d* value for the relaxation between *s* and *t* is larger than that between *s* and *b*, just as it is shown in Fig. 7.12 that at $t = 5$ ps, $d = 1.53$ for the former case while $d = 0.85$ for the latter one.

Wolynes *et al.* have considered the energy relaxation of a molecular quantum state as a random walk in the phase space. (Refs.7.6,7.7) They considered

$P(t) \sim V(t)^{-1}$ with V the space that the random walk can reach. Together with the scaling theory and Anderson localization, they inferred that as t is small, $P(t) \sim t^{-1}$ while for large t, $P(t) \sim t^{-d_f/\alpha}$. Here, d_f is the fractal dimension, and α is a constant. Our result is consistent with the analysis by Wolynes *et al.* Apparently, there exists a fractal character in the relaxation mechanism.

7.10 Action localization

For the high excitation with a large quantum number, we can explore the action localization for an eigenstate with energy E by $E = H_0(n_s, n_t, N)$. This is an important topic since action (energy) localization implies that energy is localized in the individual bonds for that state. This meets the expectation that energy can be concentrated in certain bonds and therefore excitation can lead to selective dissociation of bonds or the controlling of a chemical reaction.

From Table7.1, we know that for H_2O, D_2O, H_2S and CD_2^-, $H_0(n_s, n_t, N) - E$ is a second order polynomial in n_s, n_t. Hence, the solution (n_s, n_t) to $H_0 - E = 0$ will be parabolic, hyperbolic or elliptic. The resonance term H_r only broaden these curves. In fact, as $N \leq 20$, all (n_s, n_t) traces are elliptic.

Take the 66 levels with $N = 10$ of H_2O, as an example. Levels are labeled by L# (# =1, 2, ···) from the lowest one. Fig.7.13(b) shows the (n_s, n_t) traces of L10, L19, L60.

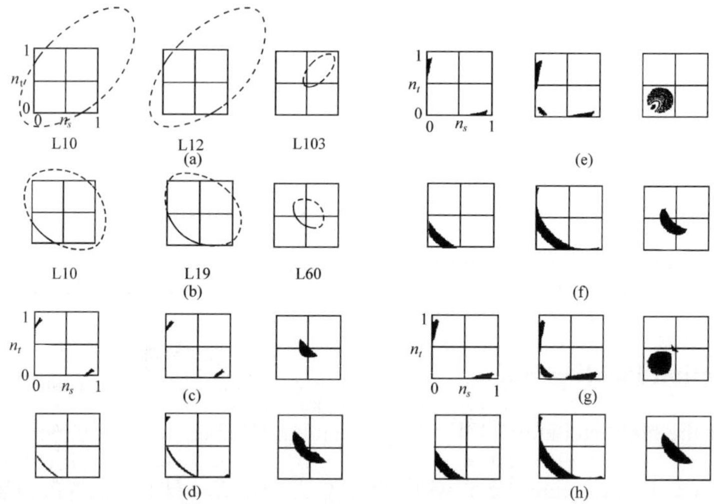

Fig.7.13 The solution (n_s,n_t) to $H_0 - E = 0$ for H_2O and CH_2- in various situations. See text for details.

In Fig. 7.13, n_s, n_t are normalized by N. Solid curves show the solutions. Fig.7.13 (d),(f) are the results for H_0 plus K_{st} and K_F, respectively. (The K_{DD} term is neglected because of its small value) Fig.7.13(h) is the result for the complete Hamiltonian $H_0 + H_r$. It is seen that the H_r, K_{st} or K_F terms only broaden the solutions to $H_0 - E = 0$. In (n_s, n_t), L10 and L60 show a connected region while L19 shows three disconnected regions. Classically, a trajectory therein will remain in the region where it starts. This is the action (energy) localization. The dynamics and the vibrational configurations corresponding to these three regions cannot be transformed to each other in a continuous way. The system, then, is not ergodic. Certainly, as the potential barrier is not too high, quantum tunneling may break the disconnection and the system may become ergodic. As the density of states is not very high and as the system dynamics is nearly classical at its high excitation, quantum tunneling will be less possible. In this situation, one is probably able to predict action localization for a state simply from $H_0(n_s, n_t) - E = 0$. Neglect of resonance terms will not affect the conclusion. In other words, anharmonicity dominates action localization. The calculation shows that for H_2O, D_2O, H_2S and CD_2-, most states have a connected (n_s, n_t) region. Only few states possess three disconnected regions. Listed in Table7.2 are these states with various *N* polyad numbers.

Table 7.2 States that possess disconnected regions in (n_s, n_t) space. See text for details.

Polyad number	8	9	10	11	12	13	14	15	16
CH_2Br_2	L4,5(a)	L4–6(a)	L4,5(b) L6,7(a) L65,66(c)	L3–5(b) L6–8(a) L78(c)	L3–7(b) L8,9(a)	L3–11(b) L12(a) L103(c)	L3–15(b)	L3–18(b) L19(a) L132(c)	L3–22(b)
CD_2Br_2							L15,16	L15,16	L15–18
H_2O		L15,16	L15–20	L14–20	L13–23	L13–23	L13–27	L13–31	L13–36
D_2O			L19–21	L19–24	L19–24	L16–26	L15–27	L14–30	L13–32

Note that for H_2S the (n_s, n_t) region is always connected. Hence, H_2S is not a proper candidate for searching for action localization. From the Table, it is also asserted that CD_2- is not a choice of top priority.

The case of CH_2- is more complicated. Its (n_s, n_t) trace is not determined simply by H_0. Fig.7.13(a), (c), (e) are the results for $N = 13$, L10, L12, L103 with H_0, $H_0 + K_{st}$, $H_0 + K_F$ terms, respectively. Fig. 7.13(g) is the result with $H_0 + H_r$. From these results, it is seen that the effect by Fermi resonance is more prominent in the highly excited states. L10 possesses two disconnected regions. L12 possesses three disconnected regions. L103 possesses two disconnected regions with one larger and the other smaller. Listed in Table7.2 are those states with other *N* that possess similar (n_s, n_t) structure as L10, L12 and L103. (labeled as (a), (b), (c), respectively) Those not listed possess only one connected (n_s, n_t) structure.

In summary, for most cases, it is correct to infer that anharmonicity plays an important role in action (energy) localization.

[Note] Another possible representation is to consider two actions of bending as an identity so that the Hamilton equations of motion in Section 7.1 can be employed. For such a case, the coefficients of the algebraic Hamiltonian may vary. Parts of the work shown in this chapter have adopted this representation.

References

7. 1 W. R. Lambert, P. M. Felker and A. H. Zewail, *J. Chem. Phys.*, 79 (1981)5958.

7. 2 F. Iachello, S. Oss, *J. Mol. Spectrosc.*, 142(1990)85.

7. 3 M. K. Ahmed, B. R. Henry, *J. Phys. Chem.*, 90(1986)1081.

7. 4 O. S. Mortensen, B. R. Henry, M. A. Mohammed, *J. Chem. Phys.*, 75(1981)4800.

7. 5 V. K. Schaarschmidt, G. Gotzl, *Z. Phys. Chem.*, *(Leipzig)* 268(1987)1105.

7. 6 S. A. Schofield, P. G. Wolynes, *J. Chem. Phys.*, 98(1993)1123.

7. 7 S. A. Schofield, P. G. Wolynes, *J. Phys. Chem.*, 99(1995)2753.

7. 8 G. Wu, *Chem. Phys. Lett*, 227(1994)682.

7. 9 G. Wu and X. Ding, *Chem. Phys. Lett.*, 262(1996)421.

7. 10 G. Wu, *Chem. Phys.*Lett., 265(1997)449.

7. 11 G. Wu, *Chem. Phys. Lett.*, 270(1997)453.

7. 12 Z. Ji and G. Wu, *Chem. Phys. Lett.*, 319(2000)45.

Appendices

The fitted levels and experimental values

Table(a) H_2O

(n_s n_t n_b)	Experimental (cm^{-1})	Fitted (cm^{-1})
0 0 1	1589.50	1591.86
0 0 2	3148.20	3151.38
1 0 0	3654.50	3665.18
0 1 0	3758.30	3750.77
0 0 3	4675.80	4678.53
1 0 1	5233.80	5236.56
0 1 1	5325.60	5322.02
0 0 4	6172.00	6173.34
1 0 2	6779.40	6775.60
0 1 2	6860.50	6861.01
2 0 0	7199.00	7212.30
1 1 0	7247.00	7253.34
0 2 0	7452.70	7439.51
0 0 5	7636.40	7635.79
1 0 3	8291.40	8282.78
0 1 3	8362.90	8368.23
2 0 1	8763.50	8763.67
1 1 1	8802.60	8805.14
0 2 1	8999.00	8991.53
1 0 4	9769.60	9757.49
0 1 4	9832.20	9843.09
2 0 2	10291.40	10284.20
1 1 2	10323.00	10323.01
0 2 2	10511.60	10512.06
3 0 0	10597.90	10608.54
2 1 0	10609.80	10620.34
1 2 0	10868.00	10868.94
0 3 0	11046.50	11029.44
2 0 3	11783.00	11771.69
1 1 3	11808.20	11809.55
0 2 3	11990.30	12001.19
3 0 1	12141.10	12140.77
2 1 1	12149.50	12152.92
1 2 1	12408.40	12403.45
0 3 1	12567.40	12564.09
3 0 2	13645.30	13640.87
2 1 2	13651.10	13651.66
4 0 0	13827.80	13832.43
3 1 0	13829.30	13836.41
1 2 2	13911.90	13906.22
0 3 2	14052.90	14068.34
2 2 0	14218.20	14220.83
1 3 0	14319.30	14315.47
0 4 0	14556.00	14541.90
4 0 1	15348.40	15344.30
3 1 1	15349.30	15351.22
2 2 1	15746.60	15741.65
1 3 1	15830.10	15835.68

0	4	1	16052.00	16063.54
5	0	0	16897.30	16890.41
4	1	0	16897.50	16890.44
3	2	0	17456.30	17458.57
2	3	0	17490.70	17492.83
1	4	0	17743.00	17748.18
0	5	0	17969.90	17970.57

Table(b) D_2O

(n_s	n_t	n_b)	experimental(cm^{-1})	fitted(cm^{-1})
0	0	1	1176.10	1177.44
0	0	2	2339.20	2340.84
1	0	0	2670.80	2679.15
0	1	0	2789.80	2788.42
0	0	3	3488.60	3490.22
1	0	1	3841.60	3845.22
0	1	1	3953.90	3954.04
0	0	4	4623.50	4625.56
1	0	2	4998.40	4996.62
0	1	2	5104.10	5105.33
2	0	0	5290.30	5300.58
1	1	0	5374.30	5380.67
0	2	0	5537.70	5531.35
0	0	5	5743.30	5746.87
1	0	3	6140.30	6134.04
0	1	3	6239.80	6242.83
2	0	1	6455.30	6456.96
1	1	1	6533.00	6535.83
0	2	1	6689.60	6687.04
1	0	4	7266.70	7257.74
0	1	4	7360.10	7367.02
2	0	2	7604.90	7598.26
1	1	2	7676.50	7674.79
0	2	2	7826.80	7828.91
3	0	0	7852.10	7860.18
2	1	0	7900.70	7908.83
1	2	0	8056.20	8054.72
0	3	0	8238.30	8226.00
2	0	3	8738.60	8725.57
1	1	3	8804.40	8801.07
0	2	3	8948.50	8959.36
3	0	1	9010.00	9008.90
2	1	1	9053.10	9054.37
1	2	1	9203.70	9202.90
0	3	1	9377.30	9375.15
3	0	2	10151.20	10141.09
2	1	2	10189.30	10184.04
1	2	2	10355.40	10337.77
4	0	0	10342.90	10347.81
3	1	0	10362.70	10368.25
0	3	2	10500.80	10515.07
2	2	0	10534.40	10532.74
1	3	0	10687.60	10679.94
0	4	0	10891.50	10876.28
4	0	1	11491.60	11487.27

3	1	1	11507.90	11507.26
2	2	1	11678.50	11682.28
1	3	1	11822.10	11829.74
0	4	1	12017.40	12030.40
5	0	0	12748.10	12748.85
4	1	0	12753.20	12753.20
3	2	0	12978.90	12985.56
2	3	0	13089.20	13091.37
1	4	0	13277.20	13277.88
0	5	0	13496.80	13495.51

Table(c) H_2S

(n_s	n_t	n_b)	experimental(cm^{-1})	fitted(cm^{-1})
0	0	1	1184.10	1185.91
0	0	2	2353.60	2354.10
1	0	0	2615.40	2614.56
0	1	0	2626.40	2630.10
0	0	3	3506.20	3504.59
1	0	1	3778.00	3777.92
0	1	1	3790.90	3793.48
0	0	4	4639.80	4637.37
1	0	2	4925.20	4924.21
0	1	2	4939.30	4939.91
2	0	0	5145.30	5145.03
1	1	0	5146.90	5148.90
0	2	0	5243.50	5244.34
0	0	5	5752.90	5752.44
1	0	3	6054.60	6054.22
0	1	3	6069.50	6069.78
2	0	1	6287.40	6286.40
1	1	1	6289.50	6290.81
0	2	1	6386.20	6385.98
1	0	4	7164.30	7165.52
0	1	4	7179.60	7181.19
2	0	2	7412.90	7412.60
1	1	2	7415.20	7415.90
0	2	2	7512.00	7512.09
3	0	0	7576.30	7576.33
2	1	0	7576.40	7577.29
1	2	0	7753.00	7750.81
0	3	0	7780.80	7781.42
2	0	3	8519.30	8519.97
1	1	3	8521.90	8523.11
0	2	3	8618.50	8620.11
3	0	1	8697.40	8697.26
2	1	1	8697.60	8697.49
1	2	1	8872.60	8871.54
0	3	1	8902.90	8902.19
3	0	2	9800.60	9800.16
2	1	2	9800.80	9800.25
4	0	0	9909.90	9910.08
3	1	0	9909.90	9910.26
1	2	2	9974.70	9975.21
0	3	2	10006.40	10005.90
2	2	0	10186.80	10184.35

1	3	0	10193.80	10194.69
0	4	0	10294.10	10293.03
4	0	1	11009.40	11008.19
3	1	1	11009.50	11008.49
2	2	1	11284.70	11282.91
1	3	1	11292.60	11293.17
0	4	1	11393.40	11391.59
5	0	0	12146.00	12146.34
4	1	0	12146.00	12146.34
3	2	0	12519.00	12519.61
2	3	0	12519.80	12521.31
1	4	0	12687.30	12690.10
0	5	0	12736.40	12735.02

Table(d) CD_2Br_2

(n_s	n_t	n_b)	experimental(cm^{-1})	fitted(cm^{-1})
0	0	2	-----	2058.46
1	0	0	2195.00	2199.19
0	1	0	2313.00	2310.78
0	0	4	-----	4034.65
1	0	2	4250.00	4245.82
0	1	2	4333.00	4336.97
2	0	0	4357.00	4354.85
0	2	0	4445.00	4449.86
1	1	0	4589.00	4584.32
0	0	6	-----	5943.56
1	0	4	-----	6203.74
0	1	4	-----	6279.81
2	0	2	6387.00	6387.43
0	2	2	6463.00	6454.94
3	0	0	6483.00	6484.55
0	3	0	6541.00	6543.21
1	1	2	6584.00	6586.60
2	1	0	6679.00	6678.91
1	2	0	6829.00	6829.79

Table(e) CH_2Br_2

(n_s	n_t	n_b)	experimental(cm-1)	fitted(cm-1)
0	0	2	2774.00	2775.59
1	0	0	2990.00	2992.69
0	1	0	3065.00	3062.37
0	0	4	-----	5466.97
1	0	2	5740.00	5739.99
0	1	2	5781.00	5780.30
2	0	0	5890.00	5893.47
0	2	0	5926.00	5925.75
1	1	0	6082.00	6082.96

0	0	6	-----	8084.72
1	0	4	-----	8391.61
0	1	4	-----	8406.58
2	0	2	8596.00	8593.44
0	2	2	-----	8602.30
3	0	0	8696.00	8694.44
0	3	0	8696.00	8700.12
1	1	2	-----	8756.52
2	1	0	8906.00	8902.21
1	2	0	9026.00	9026.97

Chapter 8

Quantal effect of asymmetric molecular rotation

8.1 Introduction

The application of coset representation in molecular vibration studies is vast. Furthermore, this algorithm can provide the quantal effect. The coset representation provides a *classical analogy* for the dynamical picture, i.e., the trajectories in the quantum phase space. It describes the quantal effect due to the classical trajectories. The quantal effect realized will be complete when the Hamiltonian is the linear combination of the operators of the dynamical algebra. Otherwise, the realization is partial. Hence, the algorithm is a mean-field approximation of the quantal method. In this chapter, we will address the molecular rotation which is of such a case. The difference by the coset representation and the classical limit for which the Planck constant is set to zero is sometimes called the nonlinear effect. Correspondingly, the latter method is called the harmonic approximation.

8.2 Coset space representation of molecular rotation

The angular momentum operators along the molecule-fixed coordinates of an asymmetric molecule J_X, J_Y, J_Z satisfy the su(2) algebra (J_Y has to reverse its sign. See Ref. 8.1):

$$[J_X, J'_Y] = J_Z, \quad [J'_Y, J_Z] = J_X, \quad [J_Z, J_X] = J'_Y$$

with

$$J'_Y = -J_Y$$

The rotational Hamiltonian is

$$H_r = A_0 J_Z^2 + B_0 J_Y^2 + C_0 J_X^2$$

for which

$$A_0 > B_0 > C_0$$

J_X^2, J_Y^2, J_Z^2 in SU(2)/U(1) coset representation (for details, see Chapter 2) are:

$$J_Z^2(J\Omega) = [2(q_1^2+p_1^2)-J]^2 + \underline{2J^{-1}[J-(q_1^2+p_1^2)](q_1^2+p_1^2)}$$

$$J_X^2(J\Omega) = u + v$$

$$J_Y^2(J\Omega) = -u + v$$

$$u = (2-\underline{J^{-1}})(q_1^2-p_1^2)[J-(q_1^2+p_1^2)]$$

$$v = \frac{1}{2}\{4[J-(q_1^2+p_1^2)](q_1^2+p_1^2) + \underline{[J-(q_1^2+p_1^2)]^2 J^{-1} + (q_1^2+p_1^2)^2 J^{-1}}\}$$

In the above expressions, the underlined parts are the differences by the coset representation and the classical limit method, i.e.,

$$\langle J\Omega | J_i^2 | J\Omega\rangle - \langle J\Omega | J_i | J\Omega\rangle^2$$

From these expressions, we have

$$\langle J\Omega | J_X^2 + J_Y^2 + J_Z^2 | J\Omega\rangle = J^2 + J$$

The value by the classical limit is J^2. Hence, the quantal effect is J. We note that the results by the coset representation and quantum mechanics are identical, though J_i^2 is second order.

8.3 Quantum-classical transition

The coset space SU(2)/U(1) is a two-dimensional sphere. In polar coordinates (θ,ϕ) we have:

$$\langle J\Omega | J_Z^2 | J\Omega\rangle = J^2\cos^2\theta + \underline{(J\sin^2\theta)/2}$$

$$\langle J\Omega | J_X^2 | J\Omega\rangle = u + v$$

$$\langle J\Omega | J_Y^2 | J\Omega\rangle = -u + v$$

$$u = [(2J^2 \underline{-J})/4]\sin^2\theta\cos 2\phi$$

$$v = \{J^2\sin^2\theta + \underline{J[\cos^4(\theta/2)+\sin^4(\theta/2)]}\}/2$$

In the classical limit, those underlined parts will diminish. Then:

$$\langle J\Omega | J_Z^2 | J\Omega \rangle = J^2 \cos^2\theta$$

$$\langle J\Omega | J_X^2 | J\Omega \rangle = J^2 \sin^2\theta \cos^2\phi$$

$$\langle J\Omega | J_Y^2 | J\Omega \rangle = J^2 \sin^2\theta \sin^2\phi$$

These are the well known expressions along the molecule-fixed coordinates. Clearly, the quantal effect is well characterized in the SU(2)/U(1) coset space. As $J^2 >> J$, coset representation will reduce to that along the molecule-fixed coordinates.

However, we have to note that as $J >> 1$, under certain conditions, the terms due to the quantal effect can be larger than those by the classical method. Fig.8.1 shows the portions (the shaded parts) of this phenomenon in (q_1, p_1) space for J_X^2, J_Y^2, J_Z^2 as $J = 5, 30$.

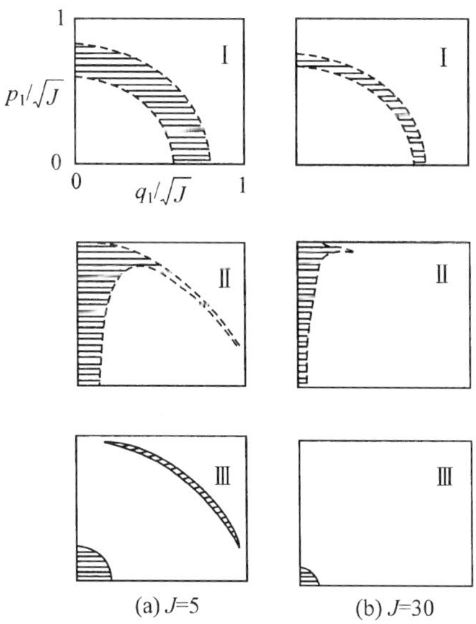

Fig.8.1 The shaded portions where the terms due to the quantal effect are larger than those by the classical method. (a) J = 5, (b) J = 30. I. J_Z^2, II. J_x^2, III. $J_x^2 + J_y^2$. The situation for J_y^2 is the same as J_x^2.

From Fig. 8.1, it is seen that, in some cases, the quantal effect is evident. As θ approaches $\pi/2$, the term by the quantal effect in J_Z^2 is larger than the classical quantity $J^2\cos^2\theta$. Similar situations also appear in J_X^2, J_Y^2 as θ approaches 0. In Fig.8.1 III one can see the situation for $J_X^2 + J_Y^2$ as $J = 5$, 30. The portions where the quantal effect is larger than the classical effect are greatly reduced. This means that as a rotor is more asymmetric, its quantal effect is more prominent. For instance, for a spherical rotor, as $J > 1$, the term due to the quantal effect is always larger than that by the classical method. Hence, for molecular rotation, not only the magnitude of J but also the degree of asymmetry will affect the accuracy of the classical treatment. This property of the rotational Hamiltonian is different from the vibration. In the vibrational Hamiltonian, the terms due to the quantal and classical effects are of the same functional forms except their orders. Hence, as action N increases, the quantal effect decreases as $1/N$. It is worthwhile to note this difference in the rovibrational dynamics.

8.4 su(2)$\otimes$h$_4$ coupling

The dynamics of a simple harmonic oscillator (or boson) can be realized by $\{a^+, a, a^+a, I\}$. This is h_4 algebra. Its subalgebras are $\{a^+a\}$ and $\{I\}$. The coset space is noncompact $H_4/U(1)\otimes U(1)$. Its structure is a two dimensional complex plane (see Section 2.5).

The Hamiltonian is

$$\omega(n+\frac{1}{2}) = \omega(z^*z+\frac{1}{2})$$

or by $z = q_2 + ip_2$,

is $$\omega(q_2^2 + p_2^2 + \frac{1}{2})$$

The rovibrational coupling is

$$\alpha(aJ_+ + a^+J_-)$$

or

$$4\alpha[J-(q_1^2+p_1^2)]^{\frac{1}{2}}(q_1q_2+p_1p_2)$$

The equations of motion are

$$\dot{q}_1 = A_0 a + B_0(-b+c) + C_0(b+c) + f$$

$$-\dot{p}_1 = A_0 + B_0(-d+e) + C_0(d+e) + g$$

$$\dot{q}_2 = 4\alpha p_1[J-(q_1^2+p_1^2)]^{\frac{1}{2}} + 2\omega p_2$$

$$-\dot{p}_2 = 4\alpha q_1[J-(q_1^2+p_1^2)]^{\frac{1}{2}} + 2\omega \dot{q}_2$$

where

$$a = (8-\underline{4/J})[2(q_1^2+p_1^2)-J]p_1$$

$$a' = (8-\underline{4/J})[2(q_1^2+p_1^2)-J]q_1$$

$$b = (2-\underline{1/J})(-2p_1)(J-2p_1^2)$$

$$c = 2p_1(2-\underline{1/J})[J-2(q_1^2+p_1^2)]$$

$$d = (2-\underline{1/J})(2q_1)(J-2q_1^2)$$

$$e - 2q_1(2-\underline{1/J})[J-2(q_1^2+p_1^2)]$$

$$f = 4\alpha\{\frac{1}{2}[J-(q_1^2+p_1^2)]^{-\frac{1}{2}}(-2p_1)(q_1q_2+p_1p_2)+[J\quad(q_1^2+p_1^2)]^{\frac{1}{2}}p_2\}$$

$$g = 4\alpha\{\frac{1}{2}[J-(q_1^2+p_1^2)]^{-\frac{1}{2}}(-2q_1)(q_1q_2+p_1p_2)+[J-(q_1^2+p_1^2)]^{\frac{1}{2}}q_2\}$$

These equations can describe the coupling of light and rotation or rovibration. The underlined parts are those due to the quantum effect.

8.5 Regular and chaotic motions

The phase space for the motion of coupled molecular rotation and a simple harmonic oscillator is (q_1, p_1, q_2, p_2). The trajectories are three dimensional in a four dimensional phase space by energy constraint. Hence, motion can be chaotic. In general, the motion with low energy is regular. Both the quantal and classical treatments lead to regular motion except their periods. Those by classical treatment possess longer periods. As energy is larger, regular and chaotic motions coexist. For the chaotic motion, quantal and classical treatments lead to irrelevant results. For a spherical rotor, no matter how large J and the coupling coefficients are, regular motion is always shown. For an asymmetric rotor with larger J, there is always chaotic motion. The relation between motion regularity and the symmetry of the rotor is obvious.

In summary, the quantal effect of rotational dynamics, especially for an asymmetric rotor, deserves attention even for large J.

References

8. 1 A. Carrington, D. H. Levy and T. A. Miller,

Adv. Chem. Phys.,18(1970)149.

8. 2 G. Wu, *Chem. Phys.*, 214(1997)15.

Chapter 9

Pendulum, resonance and molecular highly excited vibration

9.1 Pendulum

The motion of a pendulum is a very basic and important physical phenomenon. When the amplitude of a pendulum is very small, its frequency is a constant and independent of the amplitude. It is a simple harmonic motion. When its amplitude is larger, its frequency is no longer a constant and depends on the amplitude or energy. As the vibrational frequency is related to its energy, we call the system a nonlinear one.

The potential energy of a pendulum V is proportional to $1-\cos\theta = 2\sin^2\theta/2$. Here, θ is the corresponding angle. The functional form of $V(\theta)$ is shown in Fig.9.1. As θ is smaller, $V(\theta)\sim\theta^2$. It is a parabola. As θ is close to $\pm\pi$, $V(\theta)$ will not vary much. This characteristic of $V(\theta)$ leads to the quantized levels as shown in Fig.9.1. As the energy is smaller, the energy spacings are almost the same (having the character of a simple harmonic motion). When θ is close to $\pm\pi$, the energy spacing becomes smaller.

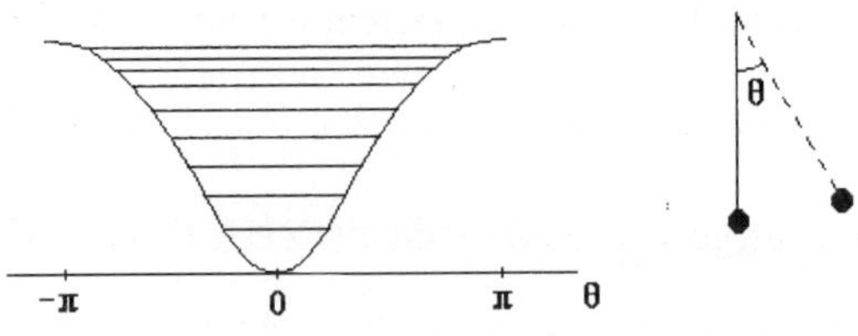

Fig.9.1 The potential of a pendulum. The horizontal lines in the potential valley show the quantized levels. Due to the nonlinear effect, the energy spacing is less for higher levels.

The phase space of the motion of a pendulum is shown in Fig.9.2. Phase space is the relation of action versus angle. (The relation of J versus θ. J is the action,

θ is the angle.) As the energy of a pendulum is not enough for θ to exceed π (or $-\pi$), the motion is stable and periodic. In this situation, the phase structure is a closed ellipse. Its center is a stable fixed point, corresponding to the stationary state. When the energy increases, θ will exceed the range of $(-\pi, \pi)$, corresponding to the rotation. The point b at the intersection of the separatrices *separating* these two types of motion is an unstable fixed point. It is a hyperbolic point. Around point b, there are both stable and unstable spaces. We note that as motion is along the separatrix, reaching to point b and stopping there, the time required is infinite. This is because . when motion is closer to point b, the speed becomes less and approaches zero!

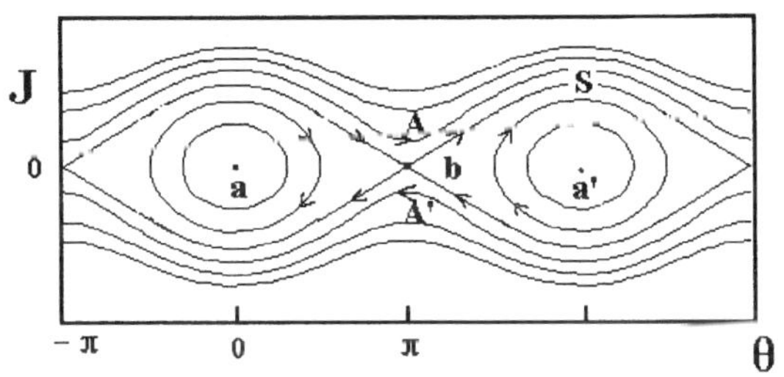

Fig.9.2 The phase diagram of the motion of a pendulum. J is the action and θ is the angle. Curves correspond to the various quantized levels. Arrows denote the directions of motion. A, A′ show the opposite rotations. S is the separatrix. a (a') and b points are stable and unstable fixed points, respectively.

9.2 Resonance

An integrable N dimensional Hamiltonian system is composed of N independent periodic motions with N conserved actions, I_i. If the frequencies of these periodic motions are proportional to integers, then the whole motion is still periodic. Otherwise, it is quasiperiodic. As there are couplings among these motions, the system becomes nonintegrable. The couplings can lead to very complicated consequences. They will destroy the previously conserved I_i. In general, we can express the couplings as the resonances of various orders. For instance, for a two dimensional system, the resonances can be 1:1, 1:2, ... etc.. The corresponding coupling terms are proportional to $\sin(\varphi_1-\varphi_2)$, $\sin(\varphi_1-2\varphi_2)$, ... etc.. (In fact, this is a Fourier expansion.)

In the following, we will show that resonance will change the structure of the phase space. Moreover, locally, it can be approximated by the motion of a pendulum!

We employ a two-dimensional system with 1:1 resonance to demonstrate this important concept.

Suppose $H_0(I_1,I_2)$ is the integrable part of the full Hamiltonian. Its frequencies are:

$$\omega_1^0=\partial H_0/\partial I_1,\quad \omega_2^0=\partial H_0/\partial I_2$$

for which $\omega_1^0\cong\omega_2^0$ or $\omega_1^0-\omega_2^0=0$

The complete Hamiltonian under resonance is:

$$H=H_0(I_1,I_2)+C_0'(I_1I_2)\sin(\varphi_1-\varphi_2)$$

Now, we make the transformations:

$$J_1=(I_1-I_2)/2,\quad \theta_1=(\varphi_1-\varphi_2)$$

$$J_2=(I_1+I_2)/2,\quad \theta_2=(\varphi_1+\varphi_2)$$

then

$$H=H_0(J_1,J_2)+C_0(J_1J_2)\sin\theta_1$$

Since H is independent of θ_2, from $\partial H/\partial\theta_2=-\dot{J}_2=0$, J_2 is a constant.

Then

$$H\cong(\partial H_0/\partial J_1)_0(J_1-J_1^0)+\frac{1}{2}(\partial^2H_0/\partial J_1^{\,2})_0(J_1-J_1^0)^2+C_0(J_1^0,J_2^0)\sin\theta_1$$

but $$(\partial H_0/\partial J_1)_0=\frac{\partial H_0}{\partial I_1}\frac{\partial I_1}{\partial J_1}+\frac{\partial H_0}{\partial I_2}\frac{\partial I_2}{\partial J_1}=\omega_1^0-\omega_2^0=0$$

hence,

$$H\sim\alpha J_1^{\,2}/2+C_0\sin\theta_1$$

with α, C_0 constants.

This is exactly the Hamiltonian of a pendulum in a potential $C_0\sin\theta_1$ with kinetic energy $\frac{1}{2}\alpha J_1^{\ 2}$. Therefore, under resonance, *locally* the phase structure is approximated by that of a pendulum. Its structure is that of Fig.9.2. We note that as there is no resonance, those curves are all straight lines due to J_1 being constant. Under the effect of resonance, they are curved. Especially, we note the birth of a couple of fixed points, one is stable (point a) and the other unstable (point b). In addition, we note that the motion with delocalized θ_1 ($-\pi \le \theta_1 \le \pi$) corresponds to weak coupling while that with localized θ_1 corresponds to strong coupling. Weak coupling between the two oscillators will result in the independence of their respective phase angles, i.e., θ_1 can extend in ($-\pi \le \theta_1 \le \pi$). For strong coupling, the phase angles will be locked in a certain range. Obviously, the levels corresponding to localized θ_1 possess higher energy than those with delocalized θ_1. This is different from the pendulum case where lower levels have smaller θ while higher levels have delocalized θ. If we make the analogy with the two O–H stretches of H_2O, then strong coupling corresponds to the normal mode and weak coupling to the local mode.

Previously, it was mentioned that the stable and unstable spaces join at point b. Suppose now that under certain perturbation, if starting from a point somewhere in the stable space close to b, the trajectory will no longer return to the starting point (Otherwise, it is a periodic trajectory!). Meanwhile, the trajectory will move slower as point b is approached. Hence, the trajectory will tangle around point b and form a chaotic region as shown in Fig.9.3.

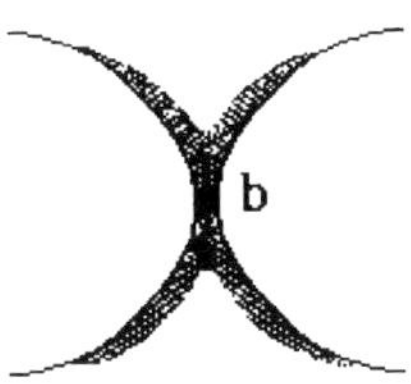

Fig.9.3 Around the unstable fixed point b, a trajectory will tangle to form a chaotic structure.

Hence, we have recognized that resonance creates chaos locally in the phase space and chaos occurs first around the unstable fixed point b.

The above analysis shows that resonance creates a stable fixed point, an unstable fixed point and the chaotic region locally in the phase space. These fixed points are periodic points. This resonance effect occurs first on the periodic trajectories. As the resonance is weak, quasiperiodic motions are relatively *stable* against resonance. They are broken down only as the resonance becomes stronger. Since the interaction in a system can be expressed as the resonances of multiple orders (as a Fourier series), all the trajectories around the stable fixed points will be broken down by the resonances of higher orders and more new stable/unstable fixed points (or periodic points) and chaotic regions will be born. This is very like the situation when we view distant mountains with our naked eyes, all we see is a green landscape. However, if we use a telescope, more and more details will show up. The processes show the character of self-similarity. Obviously, as the interaction is stronger, more periodic and quasi-periodic motions are broken down, the chaotic regions in the phase space will grow larger and finally will cover the whole phase space. This is what the noted KAM (Kolmogorov, Arnold, Moser) theorem tells us. This reveals the basic characteristics of a nonintegrable system.

The global chaos originates from the interaction among the chaos in the various parts of the phase space. This is depicted in Fig.9.4. As trajectory A from a certain resonance region enters the overlap of two resonance regions, it will be able to wander into the other region along the vertical direction to the resonance line. The trajectory will continue to wander further if it encounters more resonance overlaps. This is just like a diffusion process and will form the global chaos. That resonance overlap will result in global chaos is a criterion noted by Chirikov (See Ref. 9.1). Besides, diffuse motion can also be along the resonance line, i.e., as trajectory B in Fig.9.4. This process is called Arnold diffusion. This process has to encounter the hyperbolic point which will slow down the diffusion. Hence, its diffusion rate is slower that that of Chirikov diffusion.

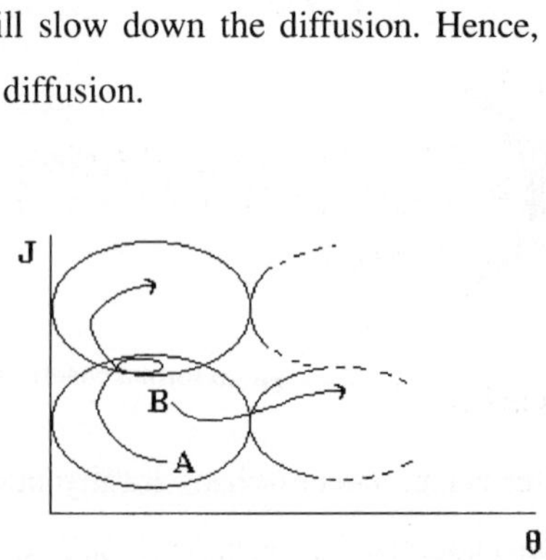

Fig.9.4 Trajectory A diffuses vertically to the resonance line (Chirikov diffusion). Trajectory B diffuses along the resonance line (Arnold diffusion).

The diffusion in the two-dimensional Hamiltonian system will be hindered by the periodic and quasiperiodic motions (the KAM curves) while for the dynamics of three and higher dimensions, this hindrance will no longer be effective. This can be easily understood in the *frequency* space (Ref.9.5). Fig.9.5(a) shows the case of two-dimensional dynamics. In Fig.9.5(a), if A is a rational point (ω_1 / ω_2 is rational.), then it corresponds to a periodic motion. Similarly, if B is an irrational point, it corresponds to a quasiperiodic motion. Region C is the chaotic region (motion) (If one prefers, chaotic motion can be considered as an admixture of motions with various frequencies.). Since motion has to be continuous, chaotic motion can only be limited to the designated region. Fig.9.5(b) shows the case of three-dimensional dynamics. Obviously, chaotic trajectories can wander unlimitedly in the phase space.

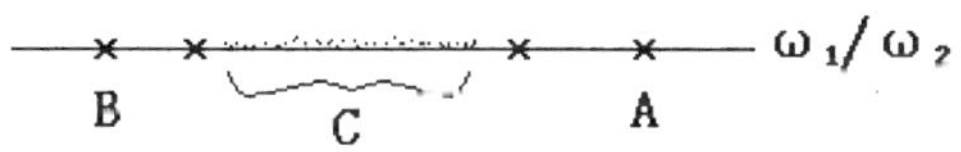

(a)

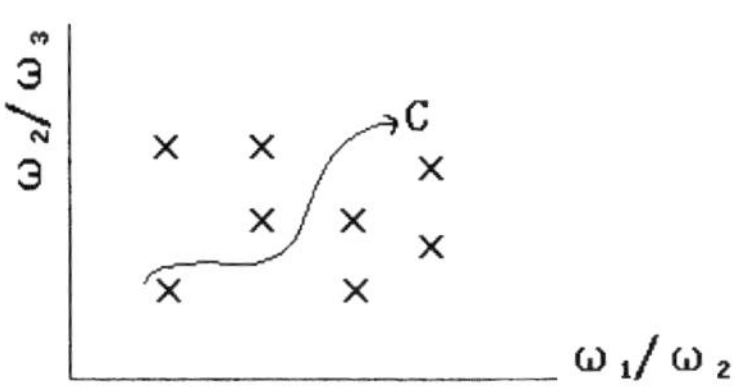

(b)

Fig.9.5 (a) One-dimensional frequency space for two-mode case. (b) Two-dimensional frequency space for three-mode case. × shows the periodic and quasiperiodic motions. C shows the chaotic motion.

9.3 Molecular highly excited vibration

The vibration of two coupled stretches can be considered as the motion of two coupled pendula. Fig.9.6 shows its potential curve in which the broken line corresponds to the separatrix. a, a′ are the stable fixed points, b is the unstable fixed point. For the levels above the separatrix, due to strong coupling, (energy can be transferred between the two oscillators) the motion will be normal and the energy spacing between the nearest neighboring levels becomes larger. The vibrational modes corresponding to a, a′ are the local modes with weak coupling in which the phase angles of the two oscillations are more or less independent without a fixed relationship.

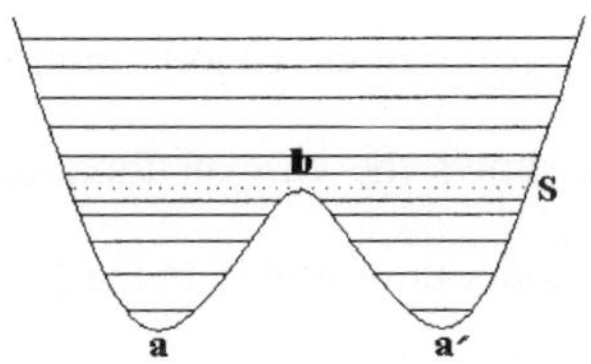

Fig.9.6 The vibrational potential of two coupled stretches. The horizontal lines are the energy levels. a, a′ are the stable fixed points. b is the unstable fixed point. S is the separatrix.

Suppose the two stretches are equivalent, the Hamiltonian is

$$H = \omega_0 (n_s+1/2)+\omega_0 (n_t+1/2)+\alpha(n_s+1/2)^2 +\alpha(n_t+1/2)^2$$
$$+\alpha_{st} (n_s+1/2)(n_t+1/2) + k\,(a_s^+ a_t+a_t^+ a_s),$$

where subscripts s, t denote the two stretches, ω_0 is the fundamental frequency (in the unit of cm^{-1}), n is the action, a and a^+ are the destruction and creation operators, α and α_{st} are the anharmonicities, and k is the coupling coefficient. Due to coupling, n_s and n_t are not conserved while $N = n_s+n_t$ is still conserved.

According to Heisenberg's viewpoint, the classical analogues of a, a^+ are:

$$a \sim (\sqrt{n})e^{-i\varphi},\ a^+ \sim (\sqrt{n})e^{i\varphi},$$

and the corresponding Hamiltonian is

$$\omega_0 (N+1)+\alpha\,(n_s+1/2)^2+\alpha\,(n_t+1/2)^2$$
$$+\alpha_{st} (n_s+1/2)\,(n_t+1/2) + 2k\,(n_s\, n_t)^{1/2} \cos\varphi_{st}\ ,$$

where φ_{st} is the phase difference. $n_s = N - n_t$.

As N is given, the eigenenergies can be obtained by diagonalizing the Hamiltonian matrix. The Hamiltonian matrix can be constructed from the bases: $\{|n_s, n_t\rangle = |0, N\rangle,\ |1, N-1\rangle,\ \ldots,\ |N-1, 1\rangle,\ |N, 0\rangle\}$. There are $N+1$ bases and there are $N+1$ eigenenergies. If we set the eigenenergy equal to the Hamiltonian, then we can obtain the relations of $n_s = n_s(\varphi_{st})$ and $n_t = n_t(\varphi_{st})$.

In this treatment of the dynamics of the two coupled stretches, the algorithm is based on su(2) algebra. In this algorithm, a conserved quantity is employed (Hence, the corresponding coset space is an S^2 sphere). The whole process is very neat, showing simple dynamical characters.

Vibration of the two coupled O–H stretches of H_2O is taken as an example. The coefficients of the Hamiltonian were elucidated from the fit to the experimental values. (See Section 7.2.) They are (in cm^{-1}) : $\omega_0 = 3890.6$, $\alpha = -82.1$, $\alpha_{st} = -13.2$ and $k = -42.7$.

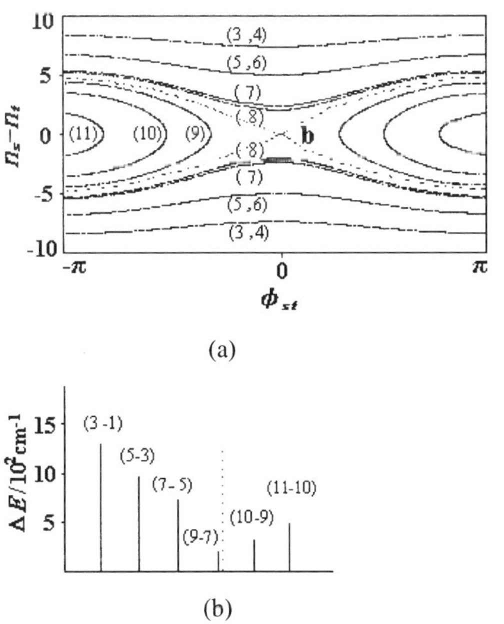

Fig.9.7 The phase diagram of the vibrational system of two coupled O-H stretches (a) and the nearest level energy spacings (b). The number in "( )" shows the level numbering. The dashed line shows the separatrix.

Fig.9.7(a) is the phase diagram ($(n_s$-$n_t)$ against φ_{st}) for the case of N = 10 (totally 11 levels) by the su(2) algorithm. It is observed that for the low levels with small coupling, the two O–H stretches possess almost the same energy and the phase angles are more or less independent and their difference, φ_{st}, can cover the whole range of $(-\pi,\pi)$. This shows the character of a local mode. At higher excitation, the coupling is stronger and the appearance of equi-energy spacings between the nearest neighboring levels is evident. Meanwhile, the phase angle difference is fixed within an interval. This is the character of a normal mode. However, there is variation from the pendulum case where small θ corresponds to low excitation while extending θ corresponds to high excitation. We also note that at the unstable fixed point b, the phase difference $\varphi_{st} = 0$. Fig.9.7(a) shows that the separatrix where the transition between the local and normal modes occurs is in between the 8th and 9th levels. Fig.9.7(b) shows that the energy spacing ΔE between the nearest neighboring levels (excluding the spacings of the states that are approximately degenerate) is the smallest around the separatrix. This is consistent with the pendulum case.

In su(2) representation, the dynamics is integrable. However, if the system is su(2) symmetry-breaking under couplings and it becomes unintegrable. Then we may infer that the motion for the levels close to the separatrix is more chaotic (stochastic) while that for the higher or lower levels is more regular. At a first glance, this seems to contradict common sense (The higher the energy, the more chaotic the motion is!). Certainly, at low excitation, the coupling is weak and the motion is naturally regular. At high excitation, stronger coupling will lead to the phase-locking phenomenon and the motion will tend to be more regular!

In conclusion, it is feasible that by the classical algebraic Hamiltonian, together with the analogy with the pendulum motion, the dynamical characters of the highly excited vibration of two coupled oscillators can be retrieved. For the case of two O–H stretches, the analysis shows that at low excitation, the coupling between the two stretches is less, the levels show approximate degeneracy and local behavior. At high excitation, due to strong coupling, the normal character with equi-energy spacings appears. For the levels close to the separatrix, analogous to the region near the unstable fixed point in the pendulum motion, chaotic motion may appear if further couplings occur so that the system becomes unintegrable. For lower and higher levels, the motion is more or less regular. Meanwhile, the energy spacings between the nearest neighboring levels close to the separatrix are less, an exact analogy with the quantized levels of the pendulum. We will discuss more on this topic in Chapter 19.

References

9. 1 B.V. Chirikov, *Phys. Rep.*, 52(1979)263.

9. 2 D. Zheng and G. Wu, *Acta Phys. Sinica*, 51(2002)53.

9. 3 M.E. Kellman, *J. Chem. Phys.* 83 (1985) 3843.

9. 4 J.P. Rose and M.E. Kellman, *J.Chem.Phys.*, 105(1996)10743

9. 5A. Morbidelli, 'Resonant Structure and Diffusion in Hamiltonian Systems' in *Chaos and Diffusion*, pp.65-112 edited by D. Benest and C. Froeschle (Cham94)

Chapter 10

Quasiperiodicity, resonance overlap and chaos

10.1 Periodic and quasiperiodic motions

For a periodic function f with period T, we have $f(t) = f(t+T)$. By Fourier transformation, its frequency function $F(\omega)$ is:

$$F(\omega) = \int_{-\infty}^{\infty} f(t)e^{i\omega t} dt = 2\pi \sum a_n \delta(\omega - n\Omega)$$

Here, $\Omega = 2\pi / T$, δ is the Dirac function, i.e., $\delta(x \neq 0) = 0$, $\delta(x = 0)$ is infinite. From this expression, we know that a periodic function possesses fundamental frequency Ω, overtone frequency 2Ω,... and n-fold overtone frequency $n\Omega$, etc..

If a system possesses N periodic motions, each with period T_i, i.e., the system possesses the function with the property:

$$g(t_1, t_2, ..., t_i, ..., t_N) = g(t_1, t_2, ..., t_i + T_i, ..., t_N)$$

and if $\Omega_i (= 2\pi / T_i)$ of these N motions are in rational ratios, then the motion of the system is still periodic. Otherwise, if the ratios are irrational, i.e., except where $\{m_1, m_2, ..., m_N\}$ are all zero, there are no integers m_i such that:

$$m_1\Omega_1 + m_2\Omega_2 + ... + m_N\Omega_N = 0$$

then the system is quasiperiodic. Its function can be written as:

$$g(t) = \sum a_{n_1 ... n_N} \exp[i(n_1\Omega_1 + n_2\Omega_2 + ... + n_N\Omega_N)t]$$

Its Fourier transformed function is

$$G(\omega) = 2\pi \sum a_{n_1 ... n_N} \delta[\omega - (n_1\Omega_1 + n_2\Omega_2 + ... + n_N\Omega_N)]$$

It is obvious that the frequency spectrum of this quasiperiodic function is

$$n_1\Omega_1 + n_2\Omega_2 + ... + n_N\Omega_N$$

Hence, the frequency spectrum of a quasiperiodic function is simply the sum of those of periodic functions. In contrast, chaotic motion possesses an irregular and continuous frequency spectrum. These are shown in Fig.10.1.

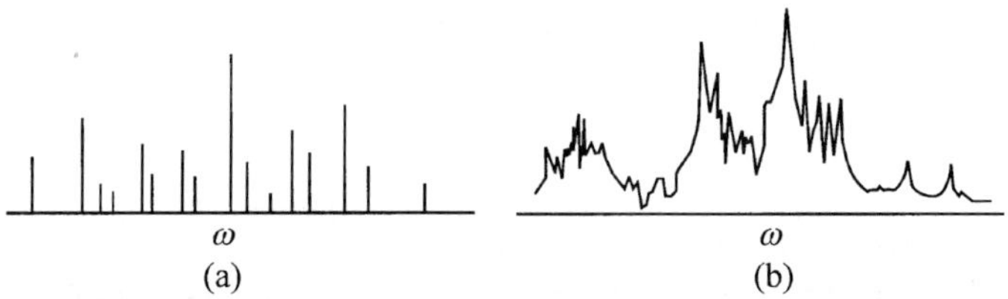

Fig.10.1 The frequency spectrum of quasiperiodic (a) and chaotic (b) motions

For a system of two frequencies (corresponding to two motions), Ω_1, Ω_2, we can use a torus as shown in Fig.10.2 to represent its dynamics. The dynamical trajectories run on the torus surface with corresponding phase angles: θ_1 and θ_2. The sizes of the torus (the size of the 'hose' and the size of the 'ring') can correspond to the actions.

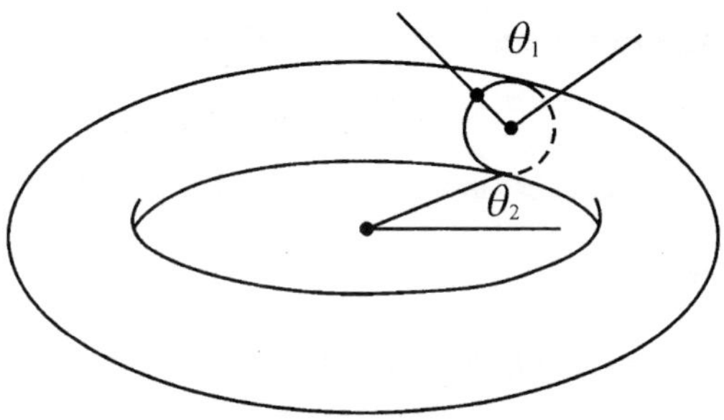

Fig.10.2 Torus for showing the dynamics of a system of two frequencies with two corresponding phase angles θ_1, θ_2.

For convenience, we define the rotational number: $R = \Omega_1 / \Omega_2$. R is the evolutional cycle of θ_1 as θ_2 evolves a cycle of 2π (See Fig.10.2).

If $R = p/q$, with p, q integers, i.e., R is rational, then at time $T = pT_1 = qT_2$, the trajectory will return to its initial position. The motion is periodic and its trajectory is a closed trace on the torus. If R is irrational, the motion is quasiperiodic and its trajectory will never return to its initial position and will cover the whole surface of the torus as time elapses.

For convenience, a surface of section with θ_2 a constant, called a Poincare surface of section, can be taken to designate the trajectory. Whenever θ_2 returns to this constant, the trajectory will go through the surface of the section and leave a point (a trace) on it. Between two consecutive points on the surface of section, we have:

$$\theta_{n+1} = \theta_n + 2\pi R$$

Here, θ_1 is shortened to θ, n shows the number of times that the trajectory pierces the surface. θ is between 0, 2π. As $R = p/q$ is rational, $\{\theta_i,\ i = 1, 2, \ldots\}$ is a finite set. Its period is q (Since $\theta_{n+q} = \theta_n + 2\pi qR = \theta_n + 2\pi p = \theta_n$). As R is irrational, $\{\theta_i\}$ will cover the whole range of $[0, 2\pi]$. The dimension of the former set is zero and 1 for the latter set.

For a multidimensional system, we can have a multidimensional torus and its multidimensional surface of section to describe the dynamics in an analogous way as shown here, though then the picture is not so intuitive.

10.2 Sine circle map

Consider that there is a nonlinear coupling of sine functional form between two periodic motions with frequencies Ω_1 and Ω_2, then the mapping for θ_i is:

$$\theta_{n+1} = \theta_n + W + k\sin\theta_n$$

Here, $W = 2\pi R$, k is the coupling coefficient.

(Refer to Section 9.3 where the coupling is of the similar form $\cos\phi_{st}$)

Under coupling, the rotational number should be averaged as:

$$\bar{R} = \frac{1}{2\pi}\lim_{m\to\infty}\frac{1}{m}\sum_{n=0}^{m-1}\Delta\theta_n$$

$$= \frac{1}{2\pi}\lim_{m\to\infty}\frac{1}{m}\sum_{n=0}^{m-1} W + k\sin\theta_n$$

If $\overline{R}$ is rational, then the motion is periodic; if $\overline{R}$ is irrational, then the motion is quasiperiodic.

This mapping of a sine functional form possesses the following properties:

(1) As $k = 0$, when $\overline{R} = R$ is in the range of [0,1], those rational points correspond to the periodic motion, while those irrational points correspond to the quasiperiodic motion. The measure of the set of rational numbers is zero while the set of irrational numbers is 1.

(2) As $k \to 0$, the measure of the set of $\{\overline{R}\}$, for which the motion is quasiperiodic, is still 1. This means that the coupling will not destroy all the quasiperiodic motion immediately.

(3) As k increases, quasiperiodic motion is destroyed gradually and forced to become periodic. As $k = 1$, the measure of the set $\{\overline{R}\}$ for which the motion is quasiperiodic is 0 while that for the periodic motion is 1. Then, except for some isolated points for which $\overline{R}$'s are irrational, there is a huge number of points whose $\overline{R}$'s are rational. In other words, most motion is periodic. The process shows that as the coupling increases, the phenomenon of 'phase-locking' is prominent with more motion becoming periodic. Fig.10.3 shows this process diagrammatically.

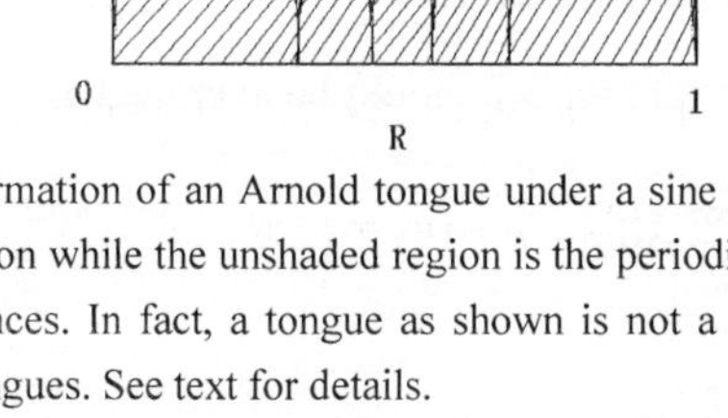

Fig.10.3 The formation of an Arnold tongue under a sine circle map. The shaded region is the quasiperiodic motion while the unshaded region is the periodic motion. The grey region shows the overlap of resonances. In fact, a tongue as shown is not a complete tongue in which there are infinitely many tongues. See text for details.

In the Figure, the shaded region is the so-called *Arnold tongue* for which $\overline{R}$ is irrational. (In fact, the tongue as shown is not a complete tongue in which there are infinitely many tongues) The unshaded region corresponds to the periodic motion with rational $\overline{R}$. This region can be called the resonance region.

(4) In the Figure, the grey region is the overlapping of resonance regions, where chaos emerges. As $k > 1$, chaotic motion is prevailing.

(5) This coupling of sine functional form shows a route to chaos. That is: starting from quasiperiodic motion, as coupling increases, motion turns out to be periodic (phase-locking effect) and finally becomes chaotic by the overlapping of resonances.

10.3 Resonance overlap and birth of chaos

That overlapping of resonances will lead to chaos is a noted conjecture by Chirikov (See Section 9.2 and Ref.10.1). Previously, Oxtoby and Rice (Ref.10.2) have also touched on the point that chaotic motion is related to the overlapping of resonances. In the following, the vibration of H_2O is employed to demonstrate this phenomenon.

Consider the two O–H stretches (labeled as s, t) and H–O–H bend (labeled as b) of H_2O vibration as anharmonic Morse oscillators with 1:1, 2:2 and Fermi resonances. In Chapter 7, it has been shown that its dynamics can be described by the SU(3)/U(2) coset space which is a three-dimensional sphere in four-dimensional space whose coordinates are (q_t, p_t, q_b, p_b). The system Hamiltonian is

$$H = H_0(q_t, p_t, q_b, p_b) + H'(q_t, p_t, q_b, p_b)$$

for which H_0 is the anharmonic Morse part and H' is the coupling part due to 1:1, 2:2 and Fermi resonances. The actions for s, t and b oscillators are $n_s = N - n_t - n_b/2$, $n_t = (q_t^2 + p_t^2)/2$, $n_b = (q_b^2 + p_b^2)/2$. N is the total action of the system and is a constant.

Resonance lines (regions) are determined by $K \cdot \omega_0 = 0$. $\omega_0 = (\omega_{0s}, \omega_{0t}, \omega_{0b})$

with $\omega_{0\alpha} = \partial H_0 / \partial n_\alpha (\alpha = s,t,b)$. For the 1:1 coupling, K =(1, -1, 0); for the 2:2 coupling, K =(2, -2, 0); for the Fermi resonance K =(1, 0, -2) and (0, 1, -2).

In addition, for a state with eigenenergy E, we can obtain its phase space from the equation $E = H(q_t, p_t, q_b, p_b)$.

Fig.10.4 shows the resonance lines for various N in (n_s, n_t) space for the systems of H_2O, CH_2- (CH_2Br_2) and CD_2- (CD_2Br_2). Solid lines are the 1:1 resonance, broken lines are the Fermi resonance.

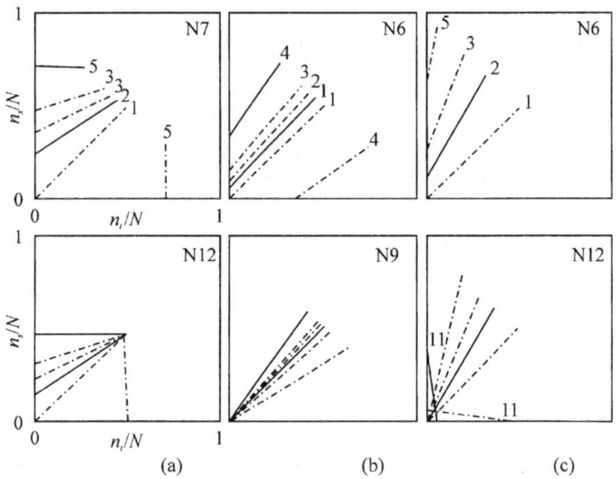

Fig.10.4 The resonance lines for (a)H_2O, (b)CH_2-, (c) CD_2- The number # in N# shows the N value. Numbers show the N values when the resonance lines first appear.

As N is small, resonance lines do not intersect. They intersect as N becomes larger. By Chirikov's conjecture, chaos will appear in the overlapping of resonances. For H_2O, as N =12, chaos will appear at the intersecting point ($n_s = n_t = N/2$) where the motion is mainly of the two stretches. As N is larger, the intersecting point moves along the diagonal line ($n_s = n_t$) toward the origin ($n_s = n_t = 0$). The dynamics is that in the highly excited vibration, those states involving mainly the bend motion will be most probably full with chaotic motion. For the systems of CH_2- and CD_2-, as N is 9 and 12, respectively, the intersection first appears at a point close to the origin. Hence, chaotic motion appears in those states involving mainly bending motion. As N is larger, the intersecting point moves along the diagonal line toward the center ($n_s = n_t =$

$N/2$). Dynamically, chaos will appear in those higher states involving more stretching.

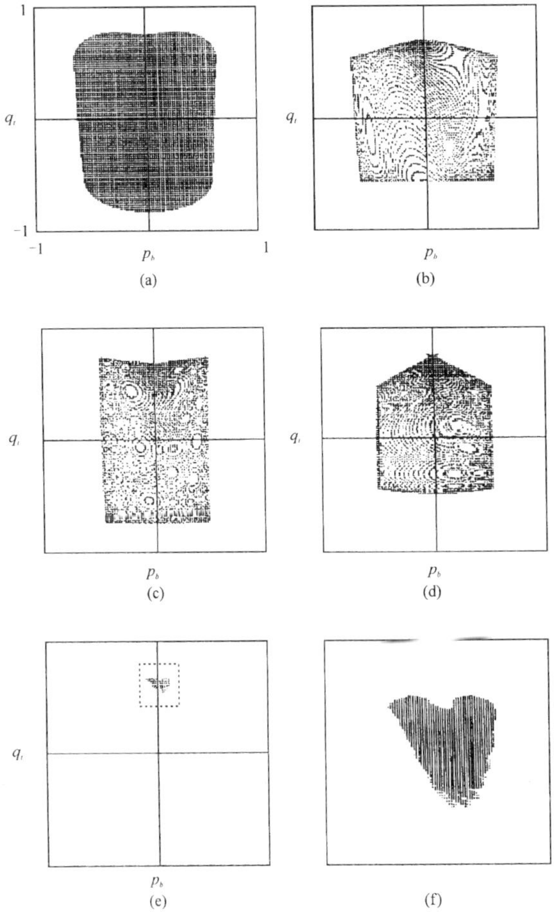

Fig.10.5 The phase space structure and resonance regions in (q_t, p_b) for the 116th state of H_2O with N = 15. (a) is the solution space by $E = H(q_t, p_t, q_b p_b)$. (b) is due to 1:1 and 2:2 resonances. (c),(d) are due to Fermi resonance. (e) is the overlap of (b), (c),(d). (f) is the enlargement of (e).

Fig.10.5 (a) is the solution space by $E = H(q_t, p_t, q_b, p_b)$. Fig. 10.5 (b) is due to 1:1 and 2:2 resonances. Figs. 10.5 (c),(d) are due to Fermi resonance. Fig. 10.5 (e) is the overlap of Figs. 10.5 (b),(c),(d). Fig. 10.5 (f) is the enlargement of Fig. 10.5 (e). From Figs. 10.5 (b)–(d), we see that resonance regions have a complicated structure, composed of many periodic, quasiperiodic motions and a small chaotic region. Also confirmed is that the overlapping region is full with chaos as conjectured by Chirikov.

In the resonance region, motion is along the K direction. Under strong coupling, resonance regions will be intertwined to form a net-like structure. Motion therein is like diffusion. This is called Arnold diffusion. It is generally recognized that this kind of diffusion plays an important role in the intramolecular vibrational relaxation (IVR) which is the theoretical background of the RRKM (Rice, Ramsperger, Kassel, Marcus) theory.

10.4 Coincidence of chaotic and barrier regions

In the nonstatistical IVR process, the barrier region in the chaotic region may be the cause of the bottleneck effect in energy relaxation. Martens (Ref.10.3) considered that this is related to the extremely irrational ratios among $\omega_{0\alpha}$. Extremely irrational numbers are those of the form $(\sqrt{5}-1)/2 \pm$ integers.

As K is extremely irrational, the corresponding trace in (n_s, n_t) space is a straight line in between the resonance lines. As N is large, the traces corresponding to irrational K will intersect with each other and the intersecting point is coincident with the intersection of the resonance lines. In other words, the extremely chaotic region contains the extreme barrier region.

Fig.10.6 shows the barrier region of the 30th level of the system of H_2O, with N = 15.

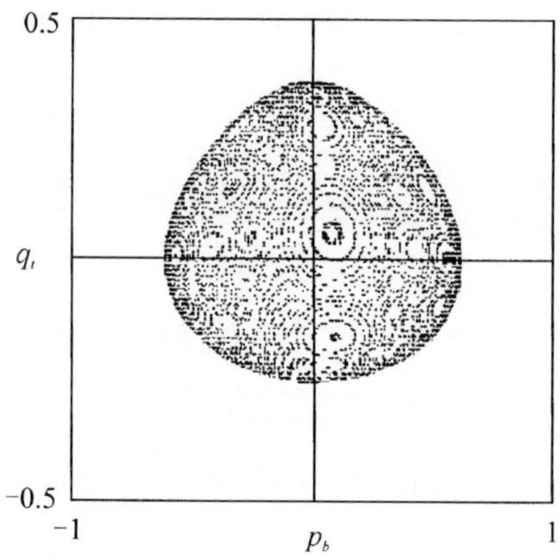

Fig.10.6 The barrier region of the 30th level of the system of H_2O, with N = 15

From Figure 10.6, it is seen that the barrier is composed of periodic and quasiperiodic motions. The periodic and quasiperiodic motions cannot deliver energy out as the chaotic motion. Hence, they stop the energy transfer.

There are two kinds of barrier region, one is the cantorus formed from the broken torus as mentioned here. The other is the broken separatrix. The chaotic region is not only occupied by the chaotic motion, but has the barrier structure. This is an important cause for the nonstatistical IVR.

References

10.1 B. V. Chirikov, *Phys. Rep.*, 52(1979)263.

10.2 D. W. Oxtoby and S. A. Rice, *J. Chem. Phys.*, 65(1976)1676.

10.3 C. C. Martens, M. J. Davis and G. S. Ezra, *Chem. Phys. Lett.*, 142(1987)519.

10.4 E. Ott, *Chaos in Dynamical Systems*, Cambridge University Press, Cambridge, 1994.

10.5 G. Wu, *Chem. Phys. Lett.*, 270(1997)453.

10.6 Z. Ji and G. Wu, *Chem. Phys. Lett.*, 319(2000)45.

Chapter 11

Fractal structure of eigencoefficients

11.1 Dimension

Measurement is the foundation of geometry and is also the core of physics. For instance, we have the measurement of space and time. Their dimensions — length, area, volume and second — are the fundamental units. Since the foundation of Euclidean geometry, over thousands of years, it has been recognized that though the shapes of objects in our world are numerous, they are the composites of simple and fundamental objects — point, line, surface and body. Any complicated objects of strange shape can be constructed from these fundamental objects. The characteristics of point, line, surface and body can be described by their dimensions.

The concept of measurement of an object is to cover the object with a unit segment of length ε, a unit square of side length ε, or a unit cube of side length ε depending on whether the object is a point, line, surface or body, respectively, and to record the number of the units required to fully cover the object. The outcome N of the measurement and ε is related by a power function:

$$N \sim \varepsilon^{-d}$$

For point, $d = 0$. For a line, surface and body, d would be 1, 2, 3, respectively. d is the dimension and it is an integer.

11.2 Fractal dimension

The concept that the dimension of a geometric object is either 0, 1, 2 or 3 has been accepted without doubt over thousands of years. But it is now being challenged.

Consider the following process: for a segment of length 1 (The exact value is not important.), we delete its middle 1/3 portion, then for the remaining two segments we delete their respective middle 1/3 portions. (See Fig.11.1) If we repeat this process infinitely, what is then the final remnant?

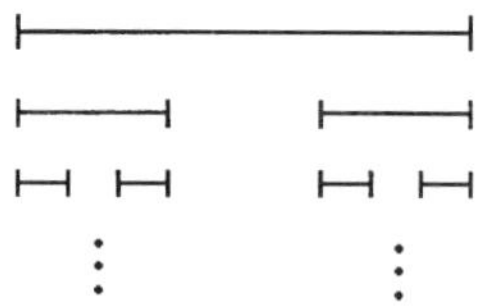

Fig.11.1 The repeated deletion of the middle 1/3 portions of all the segments in an endless way

First, we note that its length is zero. Hence, we may consider it as a set of points. This is of course right. Meanwhile, we note that in each step of deletion, a remnant point will be either on the right or left segment under deletion. Let '1' denote the case if the point is on the right segment and '0' if it is on the left segment. Then, a remnant point can be represented as $0.p_1p_2p_3\ldots$ ($p_i = 0$ or 1). Conversely, for any set of { p_1, p_2, p_3, … $p_i = 0$ or 1}, there corresponds a remnant point. Therefore, there is one to one correspondence between the remnant points and all the points in [0,1]. Hence, the remnant points cannot be in one to one correspondence to the natural numbers. They are not countable. Furthermore, we note from Fig.11.1 that there are remnant points between any two remnant points and that between any two remnant points there are deleted intervals. The set of the remnant points is both very dense and very sparse. This kind of set is called a Cantor set.

From Fig.11.1, which shows the formation of the Cantor set, we note that as the measuring unit shrinks to 1/3, the number of units required to cover the remnant set will double. In other words, we have

$$2 \sim (1/3)^{-d}$$

Hence, the dimension of the remnant set is ln2/ln3 $\approx$ 0.63. The dimension is neither 0 nor 1. This indicates that it is something between a point and a line. We call it fractal since its dimension is a fraction. If we accept the concept of fractal, we will find that fractal objects in the world are really prevailing. Point, line, surface and body are actually idealized concepts.

Here, we construct a fractal by self-similarity. By self-similarity, it is easy to construct structures of fractional dimension in one, two and three spaces. Of course, a fractal is not necessarily formed by self-similarity. For instance, it can also be formed by probability distribution. The underlying core idea is the power relation between N and ε. Furthermore, we note that a fractal may exist only in a specific range of ε. In fact, this is a general rule. The fractal idea can be extended to non-geometric realms. That is: if ε for measuring and N, the outcome, satisfy the functional form of power law, we say that it is a fractal. In this way, we generalize the fractal idea, which is originally a geometric concept, to other realms. This is a very important generalization.

Here, only the basic concept of fractal is introduced. Readers are referred to other specialized books for a more detailed description.。

A fractal's dimension can be calculated by its definition or $\lim_{\varepsilon\to 0}\frac{\ln N}{\ln 1/\varepsilon}$. It can also be calculated by self-similarity when the fractal is formed by this principle. For instance, for the above mentioned Cantor set, it is formed by two parts, the right (designated by subscript R) and left parts (designated by subscript R), after the first deletion process. For these two parts, we have:

$$N(\varepsilon) - N_L(\varepsilon) + N_R(\varepsilon)$$

These two parts are similar to their parent. Hence,

$$N_L(\varepsilon) = N\ (\varepsilon / r) \qquad N_R(\varepsilon) = N\ (\varepsilon / r) \qquad \text{(here, } r = 1/3\text{)}$$

from which we have

$$N(\varepsilon) = N(\varepsilon / r) + N(\varepsilon / r) = 2\,N(\varepsilon / r)$$

or

$$\varepsilon^{-d} = 2(\varepsilon / r)^{-d}$$

$$2\,(r)^d = 1$$

i.e., $$d = \frac{\ln 2}{\ln 1/r} = \frac{\ln 2}{\ln 3}$$

This formula can be generalized to

$$\sum_i r_i^d = 1$$

The summation runs over index i designating the self-similar components.

11.3 Multifractal

A fractal can be very complex. It can contain *inhomogeneous* parts. For such a situation, we have to consider the *weight* p_i (not only the number N) of its parts (labeled by index, *i*). The fractal dimension is now generalized to :

$$D_q = \frac{1}{1-q} \lim_{\varepsilon \to 0} \frac{\ln I(q,\varepsilon)}{\ln(1/\varepsilon)}$$

here, $I(q,\varepsilon) = \sum_i p_i^q$

For $q = 0$, this definition reduces to the previous d shown in the last section and we have $D_0 = d$. For a *homogeneous* system, $p_i = \frac{1}{N}$, then $D_q = d$ which is independent of q. This shows that q is a label of the degree of the fractal structure (different levels of structure). If D_q for various q are different, we say the object is a multifractal.

D_1 is an important parameter or dimension. It shows the information content of a system and is called the *information dimension.*

$$D_1 = \lim_{q \to 1} \frac{1}{1-q} \lim_{\varepsilon \to 0} \frac{\ln I(q,\varepsilon)}{\ln 1/\varepsilon}$$

By differentiation, $\frac{d}{dq} p_i^q \Big|_{q=1} = p_i \ln p_i$ and noting that $\sum p_i = 1$,

then

$$D_1 = \lim_{\varepsilon \to 0} \frac{\sum p_i \ln p_i}{\ln \varepsilon}$$

Besides, we note without proof an important property: $D_0 \geq D_1 \geq D_2 \geq \ldots$. In fact, q can be generalized to real numbers to describe the very delicate fine structure of a multifractal.

For a multifractal, we have

$$\sum_i p_i^q r_i^\tau = 1$$

(Note that $\sum_i r_i^d = 1$ for a simple fractal as mentioned in the last section.)

For example, for a Cantor set, we have $p_i = 1/2$, $r_i = 1/3$ ($i = 1, 2$). Hence

$$\left(\frac{1}{2}\right)^q \left(\frac{1}{3}\right)^\tau = \frac{1}{2}$$

and

$$\tau = (1-q)\ln 2/\ln 3 = (1-q)d$$

Referring to the definition of D_q, we have in general that:

$$\tau(q) = (1-q)D_q$$

It can be more convenient to use $\tau(q)$ in some cases. From the definition of D_q, we know that $\tau(q)$ is

$$\lim_{\varepsilon \to 0} \frac{\ln I(q,\varepsilon)}{\ln 1/\varepsilon}$$

By
$$D_1 = \left.\frac{\tau(q)}{1-q}\right|_{q=1}$$

then,
$$D_1 = -\left.\frac{d\tau(q)}{dq}\right|_{q=1},$$

This is a useful expression.

11.4 $f(\alpha)$ function

In this and following sections, some materials are adopted from Ref.11.1.

Besides D_q and $\tau(q)$, there is another function $f(\alpha)$ which describes the multifractal. For its introduction, we discuss the concept of distribution.

First, bisecting [0, 1] from the middle, we set a distribution $P < 1$ in [0, 0.5) and a distribution (1- P) in [0.5, 1]. Repeat the same procedure for these two parts. That is: for the left half part, we set a distribution $P < 1$ and for the right half part, a distribution (1- P) . The infinite repetition of this process will finally form a multifractal as shown below. Fig.11.2 shows this process (Ref. 11.1).

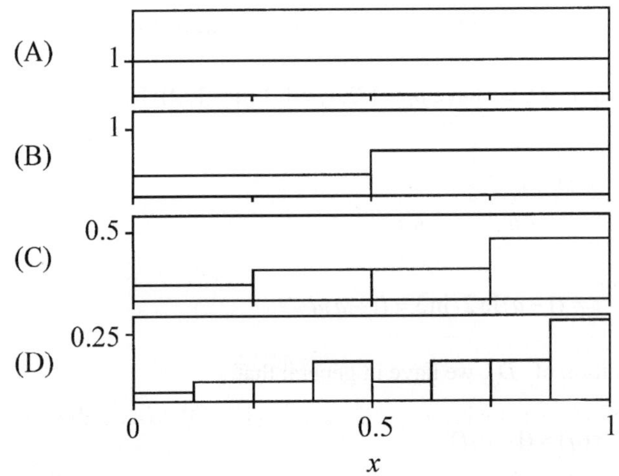

Fig. 11.2 The formation of a mulifractal by repeated distributions of P and (1-P) to every right and left part, respectively. (Adopted from Fig.1 of Chapter 9 of *Fractals, Chaos, Power Laws* by M. Schroeder, Freeman, New York 1991)

A point in [0, 1] can be labelled 1 if it is on the right portion of the line segment and 0 if it is on the left portion of the segment in each step of assigning distribution. For instance, the point at 1/5, in the first six steps, it is on the left, left, right, right, left and left portions of the respective segments. Hence, we may use 0.001100…to label 1/5. Therefore, each point is associated with a set of binary coded digits. At the sixth step, the segment where 1/5 stays is between 12/64 (0.001100) and 13/64 (0.001101) with a length of 2^{-6} and distribution $P^4(1-P)^2$. Furthermore, there are $\binom{6}{2}=15$ segments with this distribution. This can be seen from there being two 1's in the first 6 digits. With this generalization, we know that in the nth step, if there are k 1's in the first n digits, then the distribution of the segment where the point resides is $P^{n-k}(1-P)^k$, and there are $\binom{n}{k}$ such segments. The width of the segments is 2^{-n}. We define the strength (area) of these segments possessing this distribution as:

$$\binom{n}{k}P^{n-k}(1-P)^k 2^{-n}$$

(Here, we note that the total strength or area is 1 since $\sum_{k=0}^{n}\binom{n}{k}P^{n-k}(1-P)^{k}2^{-n}=1$).

By Stirling formula and defining the *H* function:

$$\binom{n}{k}\sim\left(\frac{k}{n}\right)^{-k}\left(1-\frac{k}{n}\right)^{k-n}\equiv 2^{nH(k/n)}$$

Let $\xi=\frac{k}{n}$, then

$$H(\xi)=-\xi\log_2\xi-(1-\xi)\log_2(1-\xi)$$

In general, the set of segments with distribution $P^{n-k}(1-P)^k$ and width r^n, possesses the fractal dimension $d(\xi)$:

$$d(\xi)=\lim_{n\to\infty}\frac{\log_2\binom{n}{k}}{n\log_2(1/r)}$$

For this case, $r=1/2$, d is exactly H. Here, we note that $H(1/2)=1$, $H(0)=H(1)=0$.

For convenience, we define α :

$$\alpha(\xi)\equiv\frac{\xi\log P+(1-\xi)\log(1-P)}{-\log r}$$

and call $f(\alpha)=d(\xi(\alpha))$ the spectrum of the multifractal. This is because different α (or ξ) possess different distributions and $f(\alpha)$ shows the corresponding fractal dimension. That is: different distributions correspond to different fractal dimensions. The fractals of different dimensions are not separated out. Instead, they are intertwined together.

In fact, $f(\alpha)$ and $\tau(q)$ can be related by Legendre transformation, i.e.,

$$f(\alpha)=\tau(q)+q\alpha$$

α and q are related by the following two relations:

$$\alpha=-d\tau/dq$$

$$q=df/d\alpha$$

$f(\alpha)$ possesses the functional form as $H(\xi)$ as shown in Fig. 11.3. (Ref.11.1) Its maximum is at $q = 0$ for which $df/d\alpha = 0$. While from $f(\alpha) = (1-q)D_q + q\alpha$, we have $f_{\max} = D_0$. Since $D_0 \le 1$, hence, $f_{\max} \le 1$.

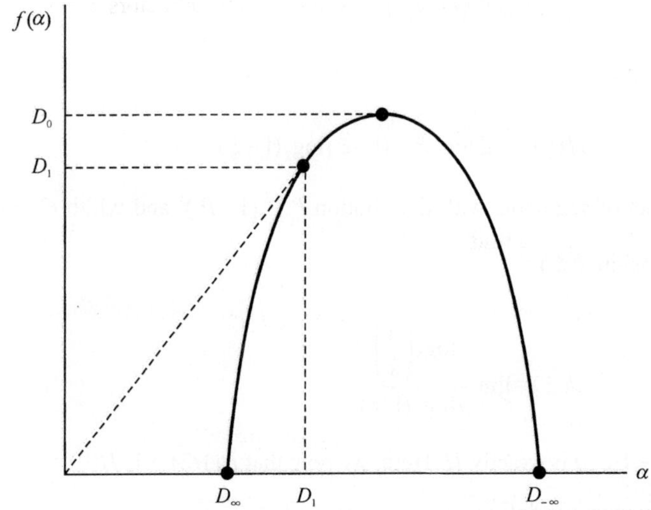

Fig.11.3 The functional form of $f(\alpha)$ (adopted from Fig.9.1(b) of *Chaos in dynamical systems* by E. Ott, Cambridge University Press, 1994)

We also note that as $q = 1$, $df/d\alpha = 1$ and $\tau(1) = 0$, $f(\alpha) = \alpha$. Meanwhile, as $q = 1$, $\alpha = D_1$ (since $-d\tau/dq|_{q=1} = D_1$) or $f(\alpha)|_{q=1} = D_1$. Denote this α as α_1, then we have $f(\alpha_1) = D_1$, as well as $df/d\alpha|_{\alpha=\alpha_1} = 1$.

From the functional form of $f(\alpha)$, we know that as $q \to \pm\infty$, $f(\alpha) \to 0$. From $(1-q)D_q = f(\alpha) - q\alpha$ we know then that $D_q = \alpha$ or $D_{-\infty} = \alpha_{\max}$ (the maximal α), $D_\infty = \alpha_{\min}$ (the minimal α) (Recall that as q is an integer, the larger the value of q, the smaller is D_q)

Shown in Fig.11.3 are the functional form of $f(\alpha)$ and D_0, D_1, D_∞, $D_{-\infty}$.

11.5 Example

In [0, 1], consider the fractal: $r_1 = 0.408$, $r_2 = r_1^2$, $P_1 = 0.5$, $P_2 = 0.5$. (Note that this is the approximation to the attractor of the famous logistic transformation $f(x) = r\,x(1-x)$ at $r = r_\infty$ (=3.5699456...). As $r < r_\infty$, the attractors are the sets of period 2^m (m = 1, 2, 3, …)).

By $r_1^d + r_2^d = 1$, we calculate d which is D_0. Since $r_1^d = (\sqrt{5}-1)/2$, hence D_0 ≈0.537.

By $p_1^q r_1^\tau + p_2^q r_2^\tau = 1$, we find

$$\tau(q) = \log[(\sqrt{2^{q+2}+1}-1)/2]/\log r_1$$

or

$$D_q = \log[(\sqrt{2^{q+2}+1}-1)/2]/[(1-q)\log r_1]$$

By $D_1 = -d\tau/dq\,|_{q=1}$, we find $D_1 = 0.515$.

Furthermore, we can calculate $D_2 = 0.497$, $D_3 = 0.482$, …

By approximation: $(1+2^{q+2})^{1/2} \cong 1+\frac{1}{2}2^{q+2}$ ($q \to -\infty$), we find $D_{-\infty} = \frac{-1}{\log_2 r_1} = 0.773 = \alpha_{max}$. As $q \to \infty$, $(1+2^{q+2})^{1/2} \cong 2^{\frac{q}{2}+1}$, we find $D_\infty = \frac{-1}{\log_2 r_1^2} = 0.387 = \alpha_{min}$.

So, this fractal is a multifractal.

11.6 Fractal of eigencoefficients

We will consider that there are fractal structures in the eigencoefficients of highly excited vibrational systems. We consider two identical Morse oscillators with 1:1 and 2:2 (Darling-Dennison) couplings. (See Section 4.1) The system Hamiltonian is :

$$\omega_0(n_1+n_2+1)+\alpha[(n_1+1/2)^2+(n_2+1/2)^2]\;+\alpha_{12}(n_1+1/2)(n_2+1/2)$$

$$+\frac{1}{2}[\beta+\varepsilon/2(n_1+n_2+1)](a_1^+a_2+a_2^+a_1)+\delta(a_1^+a_1^+a_2a_2+a_2^+a_2^+a_1a_1)$$

Here, a^+, a are the creation and destruction operators, n the vibrational quantum number, ω the eigenfrequency, α, α_{12}, β, ε, δ are anharmonic coefficients and coupling coefficients, respectively. These values can be elucidated by the fit of the eigenenergies of the algebraic Hamiltonian to the experimental values. For H_2O, they are ω_0=3871.86 cm^{-1}, $\alpha=-161.37$ cm^{-1}, $\alpha_{12}=-11.92$ cm^{-1}, $\beta=$ -111.77 cm^{-1}, $\varepsilon=10.98$ cm^{-1}, $\delta=-0.99$ cm^{-1}. For O_3 : $\omega_0=1110.67$ cm^{-1}, $\alpha=-42.31$ cm^{-1}, $\alpha_{12}=$ 11.15 cm^{-1}, $\beta=49.12$ cm^{-1}, $\varepsilon=11.52$ cm^{-1}, $\delta=-2.01$ cm^{-1}.

If we employ bases $\{\,|i\rangle_1|j\rangle_2\,\}$ to construct the Hamiltonian matrix, the eigenfunctions can be written as:

$$|\varphi\rangle=\sum C_{ij}|i\rangle_1|j\rangle_2$$

i, j are nonnegative integers. C_{ij} are eigencoefficients. For convenience, define $n=i(N+1)+j+1$ with N the maximum of i, j. With index n, $|\varphi\rangle$ can be written as :

$$|\varphi\rangle=\sum C_n|n\rangle$$

n ranges from 1 to $(N+1)^2$. From previous discussions, we know that the characteristic of a fractal is embedded in the relation between the measuring ε ($\equiv 1/L$) and the outcome $A(L)$. For the one-dimensional case, we have

$$A(L)\sim\int_0^L\rho(r)dr\sim L^{d_f}$$

here, $\rho(r)$ is the density.

To avoid origin dependence of this integral, we take:

$$\int dr_0\rho(r_0)\int dr\rho(r+r_0)$$

For eigencoefficients, we can consider:

$$|C_k|^2\rightarrow\rho(r_0),\quad |C_{i+k}|^2\rightarrow\rho(r+r_0)$$

Then we have

$$A(L) = \sum_k |C_k|^2 \sum_{i=0}^{L} |C_{i+k}|^2$$

If there is fractal in the eigencoefficients, then the plot of $\ln A(L)$ against $\ln L$ must be a linear relation with slope d_f.

For H_2O and O_3, the quantum numbers at dissociation are about 25 and 27, respectively (considered as Morse oscillators, see Section 1.2). The choice of $N = 19$ should be appropriate. We then have a Hamiltonian matrix of 400 × 400 and 400 eigenenergies.

The calculation shows that for most levels, there is no fractal in the eigencoefficients. However, there are certain levels that possess linearity in the plot of $\ln A$ against $\ln L$, i.e., the fractal structure. As an example, Fig.11.4 shows the results of the 353rd level of H_2O (42249.43 cm^{-1}) and the 330th level (15750.91 cm^{-1}) of O_3.

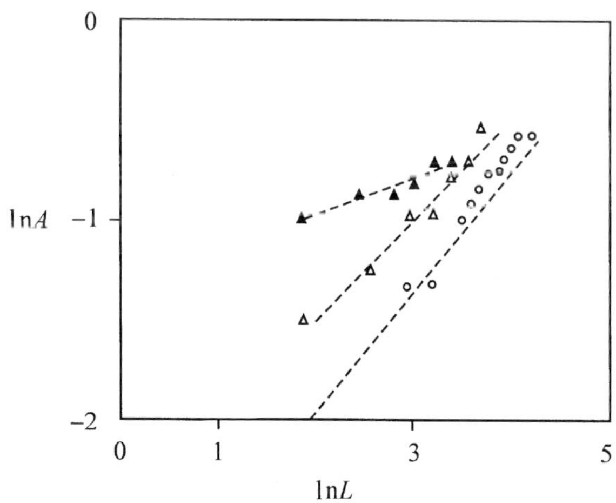

Fig.11.4 The plot of $\ln A$ against $\ln L$ of the 353rd level of H_2O (▲), the 330th level of O_3 (△) and the 260th level of H_2O with $\delta = -100$ cm^{-1} (○)

Obviously, in a certain range of L, there is a power relation between A and L. Their fractal dimensions are 0.2 and 0.5, respectively. To check this fractal character, we consider the case of stronger coupling by letting $\delta = -100$ cm^{-1}. Fig.11.4 shows the result of the 260th level (36299.20 cm^{-1}) of H_2O. The slope becomes larger with d_f

= 0.6. For this stronger coupling, we have more levels possessing this fractal character and the range of L that satisfies the linear relation in the plot of $\ln A(L)$ versus $\ln L$ is larger. Meanwhile, the distribution of C_i (against $|n\rangle$) also shows a wider extent. This property is only a necessary condition for the fractal since there are levels that show extensive distribution of C_i but do not possess a fractal character.

From this analysis, we understand that the eigencoefficients of some, but not all, highly excited states possess fractal character. The conjecture is that this fractal structure is not trivial.

11.7 Multifractal of eigencoefficients

For nonlinear triatomic molecules, their algebraic Hamiltonians can be written in the form of (see Section.7.2):

$$H = H_0 + H'$$

$$H_0 = \omega_s(n_s + n_t + 1) + \omega_b(n_b + \frac{1}{2}) + X_{ss}[(n_s + \frac{1}{2})^2 + (n_t + \frac{1}{2})^2]$$

$$+ X_{bb}(n_b + \frac{1}{2})^2 + X_{st}(n_s + \frac{1}{2})(n_t + \frac{1}{2}) + X_{sb}(n_s + n_t + 1)(n_b + \frac{1}{2})$$

and

$$H' = K_{st}(a_s^+ a_t + h.c.) + K_{DD}(a_s^+ a_s^+ a_t a_t + h.c.) + K_{sbb}(a_s^+ a_b a_b + a_t^+ a_b a_b + h.c.)$$

Here, subscripts s, t, b denote the two stretches and a bend. The rest are as shown in the last section. The coefficients are listed in Table 7.1.

For this system, $n_s + n_t + \frac{n_b}{2}$ is conserved and is denoted as N. For a given N, there are $(N + 1)(N + 2)/2$ levels. The coefficients in the Hamiltonian are determined by the fit of the level energies with $N \leq 7$ to the experimental values. In the following, we will deal with the situations with $N = 20$, $N = 30$. Then the coefficients could be in variation with those listed in Table 7.1. This issue can only be resolved by more experimental values available in the future. However, we note that in our discussion of the fractal structure, the general conclusions should not be so crucially dependent on the specific values of these coefficients. What we care about are the general concepts or conjectures.

A highly excited state $|\varphi\rangle$ can be written as

$$|\varphi\rangle = \sum C_i |\phi_i\rangle$$

$|\phi_i\rangle$ is $|n_s, n_t, n_b\rangle$ which is an eigenstate of H_0. $|\phi_i\rangle$'s are arranged in a sequence beginning from the one of the lowest energy. We arbitrarily call this the *energy coordinate*.

From the relation shown in Section 11.2:

$$\sum p_i^q r_i^\tau = 1$$

let $r_i = \delta$, then we have

$$\sum p_i^q = \delta^{-\tau}$$

We partition the range in the energy coordinate into intervals with width δ and label them by index k. In each interval, the corresponding $|\phi_i\rangle$'s are labeled by subscript j from the lowest to the highest energy. Now, consider the correspondence:

$$\sum p_i^q \to \sum_k \sum_j | C_{kj} |^2 \equiv \chi(q)$$

Hence, if there is a linear relation in the plot of $\ln \chi$ against $\ln \delta^{-1}$, its slope is $\tau(q)$. Then, the multifractal character is established. From $\tau(q)$, other multifractal quantities like $D(q)$ (D_q), $\alpha(q)$, $f(q)$, $f(\alpha)$ are easily obtained.

In the calculation, the range of δ is from 400 to 1000 cm^{-1}. To avoid numerical overflow, for $q < 0$, as $| C_i |^2 < 10^{-12}$, C_i is set to 0. The linearity of the plot of $\ln \chi$ versus $\ln \delta^{-1}$ is recognized when its correlation coefficient is larger than 0.95 (> 0.90 as q approaches ± 20).

Fig.11.5 shows the 496 energy levels with $N = 30$, of H_2O, D_2O and H_2S, among which there are 176 levels of H_2O, 143 levels of D_2O and 159 levels of H_2S that show multifractal structure in their eigencoefficients. These levels are of higher energy and high density of states.

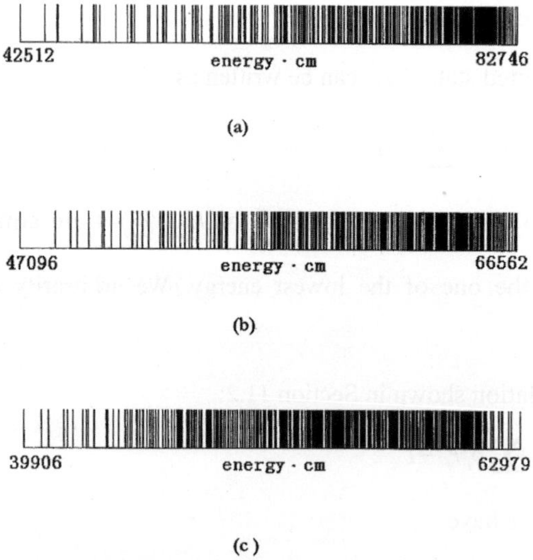

Fig.11.5 The 496 energy levels with $N = 30$ of (a)H_2O, (b)D_2O and (c)H_2S

Fig.11.6 shows the rich multifractal properties of the 194th level of H_2O with $N = 20$. Fig.(a) shows the case for $q = 2$. The slope is $\tau(2) = -0.724$ with linear correlation 0.984.

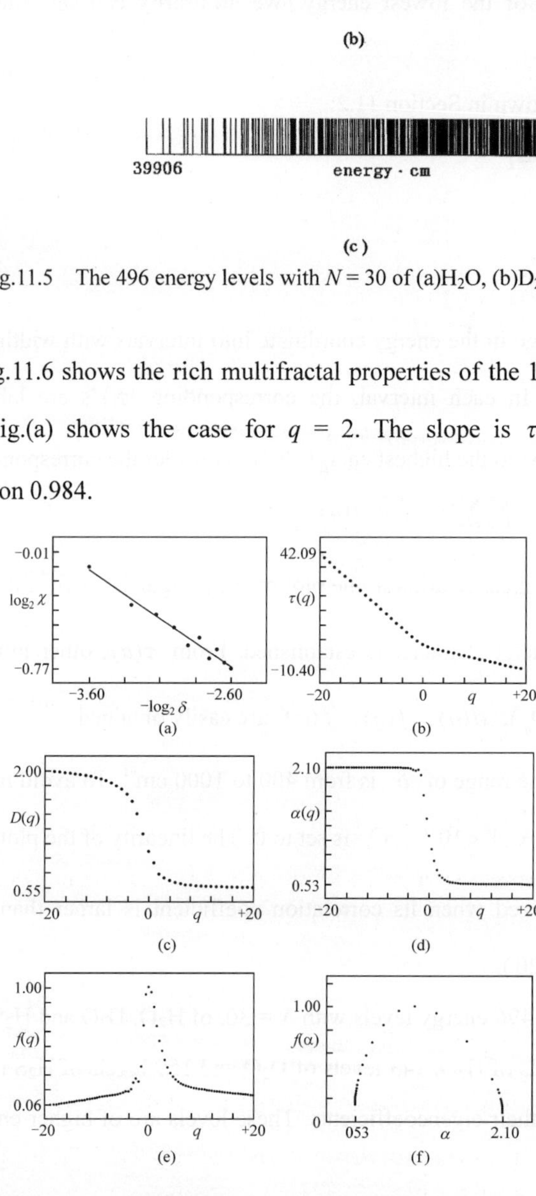

Fig.11.6 The multifractal properties of the 194th level of H_2O with $N = 20$.

There will not be one to one correspondence according to the sequence in energy between the levels of H_0 and $H_0 + H^{'}$, *adiabatically*, when they are arranged in energy coordinates from low to high energy. That is, the mth level of H_0 will not correspond exactly to the mth level of $H_0 + H^{'}$, adiabatically. However, in a rough sense for such a huge number of levels, we can still label the mth level of $H_0 + H^{'}$ by the quantum numbers (n_s, n_t, n_b) of the mth level of H_0. Fig.11.7 shows the distribution of (n_s, n_t, n_b) of the levels that possess the multifractal structure of H_2O with $N = 30$.

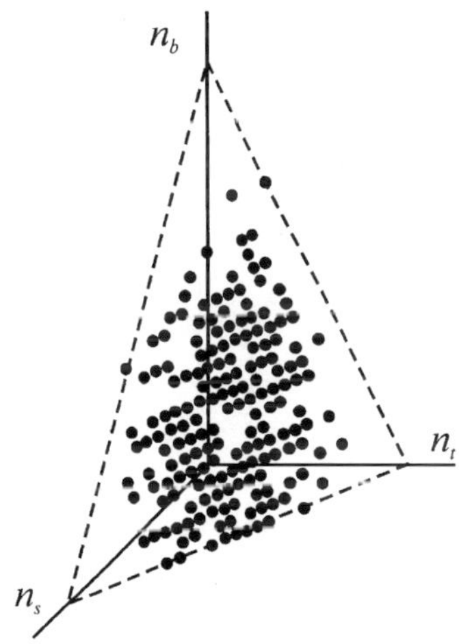

Fig.11.7 The distribution of (n_s, n_t, n_b) of the levels that possess the multifractal structure of H_2O with $N = 30$

From the Figure, it is seen that in action space, those levels that possess multifractal structure are in the interior region. In addition, the calculation also indicates that as N is larger, there are more levels that possess multifractal structure.

11.8 Self-similarity of cigencoefficients

Fractal structure is often connected with self-similarity. Fig.11.8 shows the self-similarity of the C_i distribution of the 157th level of H_2S with $N = 30$. There is a multifractal structure for this level. The dashed curves are for better visualization of the self-similarity.

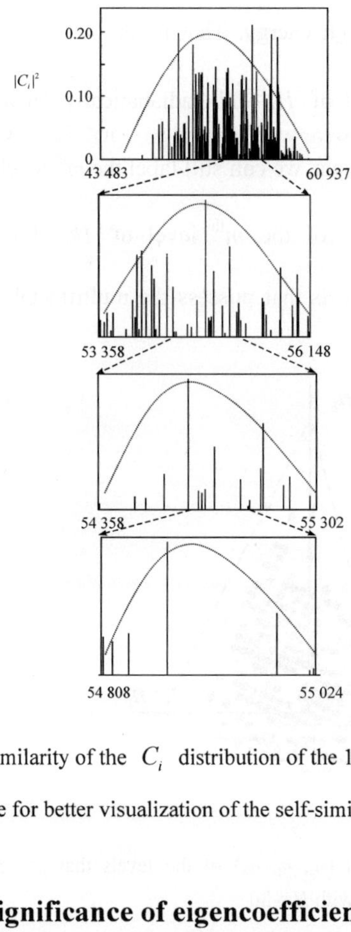

Fig.11.8 The self-similarity of the C_i distribution of the 157th level of H_2S with $N = 30$. The dash curves are for better visualization of the self-similarity.

11.9 Fractal significance of eigencoefficients

An eigenstate can be written as the linear combination of basis functions. Linear combination coefficients show the probabilities that the eigenstate is found in the basis functions. The analysis shows that in addition to probability, the coefficients also possess fractal structure and self-similarity. These are geometric characteristics which appear at high excitation as collective and global properties and are distinct from probability, which is significant only for the individual coefficient. This is a geometric indication, as the system approaches the classical limit, of the collection of eigencoefficients which are quantal in origin. This geometric character of wavefunction in the classical limit may deepen our understanding of quantum mechanics.

References

11. 1 M. Schroeder, *Fractals, Chaos, Power Laws*, Freeman, New York 1991

11. 2 E. Ott, *Chaos in Dynamical Systems*, Cambridge Univ. Press, Cambridge, 1994.

11. 3 G. Wu, *Chem. Phys. Lett.*, 242(1995)333.

11. 4 J. Yu, S. Li and G. Wu, *Chem. Phys. Lett.*, 301(1999)217.

Chapter 12

C-H bend motion of acetylene

12.1 Introduction

Acetylene is a prototype of organic compounds. It is also an important material of energy. It has two carbon atoms and two hydrogen atoms. The C–H bond is single and the C–C bond is triple. The H atom will migrate to another C atom to form vinylidene if its amplitude of bend motion is large enough. This process can be modeled as a 1,2-shift in organic chemical reactions. One way to reach high bend motion is through light absorption. However, this is impractical due to relaxation. The SEP (stimulated emission pumping) and dispersed fluorescence technique are first used to excite molecules to higher electronic states (X state), then as the molecules relax (either via stimulated emission or fluorescence) to the electronic ground states, their vibrational or rotational states can be in high excitation. The spectra are generally very complicated and their recognition and assignment are not easy tasks.

Acetylene has five modes: symmetric C–H stretch (labeled as 1), C–C stretch (2), antisymmetric C–H stretch (3), *trans* C–H bend (4) and *cis* bend (5). SEP and dispersed fluorescence experiments show that the highly excited vibrational states obtained are via the couplings of Darling–Dennison (DD) and vibrational l doubling due to the C–H bend. (For simplicity, we use l to denote both the angular momentum and its quantum number). In the early stage of the formation of these highly excited states, quantum numbers $N_{str} = n_1 + n_2 + n_3$, $N_{res} = 5n_1 + 3n_2 + 5n_3 + n_4 + n_5$ and $l = l_4 + l_5$ are conserved under these resonances. These conserved quantum numbers are called the polyad numbers. (Note that n, N, l are notations for quantum numbers)

Among these highly excited states, those with $N_{str} = 0$ corresponding to a pure C–H bend are of particular interest. This is related to the topic of 1,2 migration. For this case, $N_{res} = n_4+n_5 \equiv N_b$ is conserved. Currently, the C–H bend state up to 15000 cm^{-1} has been assigned for which N_b is close to 22. This energy is approximate to that required for H migration which is about 48 kcal/mol. Equivalently, n_4 or n_5 is 23–28.

12.2 Empirical C–H bend Hamiltonian

From the formation mechanism of the C–H highly excited bend motion (as stated in the last section), the empirical Hamiltonian H_{eff} can be written with its coefficients determined from the fit to the experimental data. They are listed in Table 12.1 (Ref.12.1)

Table 12.1 The coefficients of the empirical Hamiltonian in cm^{-1}

ω_4	ω_5	x_{44}	x_{45}	x_{55}	g_{44}	g_{45}	g_{55}
608.656	729.137	3.483	−2.256	−2.389	0.676	6.671	3.535
y_{444}	y_{445}	y_{455}	y_{555}	r^0_{45}	r_{445}	r_{545}	s_{45}
−0.0306	0.0241	0.0072	0.00954	−6.193	0.0303	0.0109	−8.572

H_{eff} has the explicit form:

(1) The diagonal elements:

$$\langle n_4 l_4, n_5 l_5 | H_{eff} | n_4 l_4, n_5 l_5 \rangle$$
$$= \omega_4 n_4 + \omega_5 n_5 + x_{44} n_4^2 + x_{55} n_5^2 + x_{45} n_4 n_5 + g_{44} l_4^2 + g_{45} l_4 l_5 + g_{55} l_5^2$$

Here, ω shows the mode frequency in cm^{-1}. x, g are nonlinear coefficients.

(2) The off-diagonal terms:

(a) DD-I:

$$\langle n_4 l_4, n_5 l_5 | H_{eff} | (n_4\text{-}2) l_4, (n_5+2) l_5 \rangle$$
$$= s_{45}/4\ [(n_4^2 - l_4^2)(n_5 + l_5 + 2)(n_5 - l_5 + 2)]^{1/2}$$

(b) DD-II

$$\langle n_4 l_4, n_5 l_5 | H_{eff} | (n_4\text{-}2)(l_4 \mp 2), (n_5+2)(l_5 \pm 2) \rangle$$
$$= (r_{45} + 2g_{45})/16[(n_4 \pm l_4)(n_4 \pm l_4 - 2)(n_5 \pm l_5 + 2)(n_5 \pm l_5 + 4)]^{1/2}$$

(c) vibrational l doubling:

$$\langle n_4 l_4, n_5 l_5 | H_{eff} | n_4 (l_4 \pm 2), n_5 (l_5 \mp 2) \rangle$$
$$= r_{45}/4\ [(n_4 \mp l_4)(n_4 \pm l_4 + 2)(n_5 \pm l_5)(n_5 \mp l_5 + 2)]^{1/2}$$

Here, $r_{45} = r_{45}^{0} + r_{445}(n_4\text{-}1) + r_{545}(n_5\text{-}1)$

We note that H_{eff} is not of analytical form and is parametrized. In fact, H_{eff} possesses a very compact and simple analytical form which corresponds to two coupled su(2) algebras as shown below.

12.3 Second quantization representation of H_{eff}

Rewrite n_4, l_4, n_5, l_5 as:

$$n_4 = n_{4+} + n_{4-},\ n_5 = n_{5+} + n_{5-}$$

$$l_4 = n_{4+} - n_{4-},\ l_5 = n_{5+} - n_{5-}$$

The subscripts "+", "-" can be understood as the right and left rotational motions. The relation between n and l is:

$$l_i = n_i, n_i\text{-}2, \ldots, -n_i, \qquad i = 4, 5$$

The conserved l, N_b are:

$$l = l_4 + l_5 = n_{4+} + n_{5+} - n_{4-} - n_{5-}$$

$$N_b = n_4 + n_5 = n_{4+} + n_{5+} + n_{4-} + n_{5-}$$

(Note that for H_{eff} in the last section, l and N_b are indeed conserved.)

For the "+" and "-" motions, we have the constants:

$$P_1 \equiv (N_b + l)/2 = n_{4+} + n_{5+}$$

$$P_2 \equiv (N_b - l)/2 = n_{4-} + n_{5-}$$

This shows that the quantum numbers for "+" and "-" motions are conserved.

In the new variables (n_{4+}, n_{4-}, n_{5+}, n_{5-}), the diagonal elements of H_{eff} are trivial. However, the off-diagonal elements are very compact. For instance, DD-I changes to:

$$s_{45}\,[n_{4+}\, n_{4-}\,(n_{5+}+1)(n_{5-}+1)]^{1/2}$$

which is:

$$s_{45}\,(a_{4+}^{+}\, a_{4-}^{+}\, a_{5+}\, a_{5-} + h.c.)$$

in terms of the second quantized operators. Similarly, for DD-II and vibrational l doubling, we have :

$$(r_{45}+2g_{45})/4(a_{4+}^{+}\, a_{4+}^{+}\, a_{5+}\, a_{5+} + h.c.)$$

$$(r_{45}+2g_{45})/4(a_{4-}^{+}\, a_{4-}^{+}\, a_{5-}\, a_{5-} + h.c.)$$

and

$$r_{45}(a_{4-}^{+} a_{4+} a_{5+}^{+} a_{5-} + h.c.)$$

12.4 su(2)⊗su(2) represented C-H bend motion

For "+" (the same for "-" bend) bend, we can define:

$$J_{x+} = (a_{4+}^{+} a_{5+} + a_{5+}^{+} a_{4+})/2$$

$$J_{y+} = -i(a_{4+}^{+} a_{5+} - a_{5+}^{+} a_{4+})/2$$

$$J_{z+} = (n_{4+} - n_{5+})/2$$

It is easy to show that $\{J_{x+}, J_{y+}, J_{z+}\}$ satisfy su(2) algebra (The algebra that the angular momenta satisfy). By these operators, DD-I and vibrational l doubling can be combined to become a very simple term:

$$4s_{45}J_{x+}J_{x-}$$

Here, we approximate $r_{45} \approx s_{45}$ since their values, -6.19 cm^{-1} and -8.57 cm^{-1}, are indeed very close.

Meanwhile, DD-II can be written as:

$$(r_{45} + 2g_{45})[(J_{x+}^{2} + J_{x-}^{2}) - (n_4 + n_5)/4 - (n_4 n_5 + l_4 l_5)/4]$$

We note that the last two terms can be grouped into the diagonal elements

For modeling, we can simplify the diagonal elements (though this is not necessary) as :

(1) $x_{45} n_4 n_5 + g_{45} l_4 l_5 - (r_{45} + 2g_{45}) (n_4 n_5 + l_4 l_5)/4$

simplified to

$$x_1(n_{4+} n_{5+} + n_{4-} n_{5-})$$

(2) $x_{44} n_4^2 + x_{55} n_5^2 + g_{44} l_4^2 + g_{55} l_5^2$

simplified to

$$x_2 (n_{4+}^2 + n_{4-}^2 + n_{5+}^2 + n_{5-}^2)$$

x_1 and x_2 are parameters.

(3) $\omega_4 n_4 + \omega_5 n_5 - (r_{45} + 2g_{45})(n_4 + n_5)/4$

simplified to

$$\omega_4 (n_{4+} + n_{4-}) + \omega_5 (n_{5+} + n_{5-})$$

since $(r_{45} + 2g_{45}) << \omega_4, \omega_5$, so the corresponding term can be omitted.

In summary, we have a very compact algebraic Hamiltonian $H_{\text{algebraic}}$

$$\omega_4 (n_{4+} + n_{4-}) + \omega_5 (n_{5+} + n_{5-}) + x_1 (n_{4+} n_{5+} + n_{4-} n_{5-}) + x_2 (n_{4+}^2 + n_{4-}^2 + n_{5+}^2 + n_{5-}^2)$$

$$+4s_{45} J_{x+} J_{x-} + (r_{45} + 2g_{45}) (J_{x+}^2 + J_{x-}^2)$$

This Hamiltonian is built up by two coupled "+" and "-" su(2) algebras.

$J_{x+} J_{x-}$ and $J_{x+}^2 + J_{x-}^2$ are of second order. From a symmetry consideration, there is one first order term λ $(J_{x+} + J_{x-})$. As the quantum numbers are small, this term is important. At high excitation, it is not as prominent as the second order terms.

It is interesting to note that "+" and "-" motions are coupled via the nonlinear term $J_{x+} J_{x-}$. The term $J_{x+}^2 + J_{x-}^2$ (or DD-II coupling) will not couple "+" and "-" motions, neither the first order term $J_{x+} + J_{x-}$.

12.5 Coset space representation

The modeled $H_{\text{algebraic}}$ is to show the algebraic property of H_{eff}. In practical application by the coset space representation, it is not really necessary.

In the two "+", "-" coset space SU(2)/U(1), (See Chapters 2 and 4) we have coordinates (q_+, p_+, q_-, p_-) and

$$n_{4+} = 2(q_+^2 + p_+^2),\ n_{5+} = P_1 - n_{4+}$$

$$n_{4-} = 2(q_-^2 + p_-^2),\ n_{5-} = P_2 - n_{4-}$$

The diagonal terms of H_{eff} can be easily expressed in terms of (q_+, p_+, q_-, p_-). The off-diagonal terms can be derived from the algebraic expressions (Section 12.3) as:

DD-I: $4s_{45} (n_{5+} n_{5-})^{1/2} (q_+ q_- - p_+ p_-)$

DD-II: $(r_{45} + 2g_{45})[n_{5+} (q_+^2 - p_+^2) + n_{5-} (q_-^2 - p_-^2)]$

vibrational l coupling: $4r_{45} (n_{5+} n_{5-})^{1/2} (q_+ q_- + p_+ p_-)$

Then, we have the (q_+, p_+, q_-, p_-) represented H_{eff}.

The Hamilton's equations of motion are:

$$\partial H_{\text{eff}} / \partial q_\alpha = -\mathrm{d}p_\alpha / \mathrm{d}t$$

$$\partial H_{\mathrm{eff}}/\partial p_\alpha = \mathrm{d}q_\alpha/\mathrm{d}t \qquad (\alpha = +, -)$$

The dynamics then is fully described by the trajectories in the two coupled SU(2)/U(1) spaces. Each SU(2)/U(1) space is a two-dimensional sphere in three-dimensional space.

Analogously, we may have $H_{\mathrm{algebraic}}(q_+, p_+, q_-, p_-)$.

12.6 Dynamics

$H_{\mathrm{eff}}(q_+, p_+, q_-, p_-)$ (or $H_{\mathrm{algebraic}}(q_+, p_+, q_-, p_-)$) is classical. It cannot offer the quantized energies. In the range that (q_+, p_+, q_-, p_-) are allowed:

$$P_1 \geq 2\,(q_+^2 + p_+^2) \ \geq 0$$

$$P_2 \geq 2\,(q_-^2 + p_-^2) \ \geq 0$$

the corresponding H_{eff} values just cover the whole range of quantized energies. For instance, for $N_b = 22$, $l = 8$, there are $(P_1+1)(P_2+1) = [(N_b + l)/2+1] \cdot [(N_b - l)/2+1] = 128$ levels. The energy range of these 128 levels corresponds to that covered by H_{eff}. Though $H_{\mathrm{eff}}(q_+, p_+, q_-, p_-)$ cannot offer the quantized energies, if we are only interested in the dynamical property as a function of energy, then the continuous energy offered by $H_{\mathrm{eff}}(q_+, p_+, q_-, p_-)$ is just enough. In fact, at high excitation, the density of states is very high and the concept of energy quantization may not be so crucial. The issue of how to obtain quantized energies from $H_{\mathrm{eff}}(q_+, p_+, q_-, p_-)$, i.e., the problem of quantization from $H_{\mathrm{eff}}(q_+, p_+, q_-, p_-)$ will be discussed in Chapter 17.

Fig.12.1 shows the energy range by $H_{\mathrm{eff}}(q_+, p_+, q_-, p_-)$ with $\omega_4 = 608.66\ \mathrm{cm}^{-1}$, $\omega_5 = 729.14\ \mathrm{cm}^{-1}$, $x_1 = x_2 = -3.0\ \mathrm{cm}^{-1}$, $4s_{45} = -34.28\ \mathrm{cm}^{-1}$, $r_{45}+2g_{45} = 7.14\ \mathrm{cm}^{-1}$, $N_b = 10, 16, 22$ and $l = 0, 8$. We partition the range into 5 regions and denote them by various notions as shown therein.

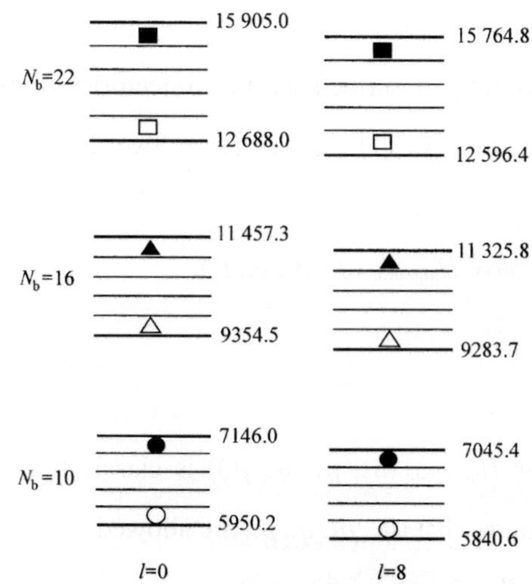

Fig.12.1 The energy range by the algebraic $H_{eff}(q_+, p_+, q_-, p_-)$ and their partitions

For a given energy (which can be eigenenergy or not) E, we can obtain the solution space (q_+, p_+, q_-, p_-) from $E = H_{algebraic}(q_+, p_+, q_-, p_-)$ (or $H_{eff}(q_+, p_+, q_-, p_-)$). The solution space is the phase space.

The action transfer between C–H *trans* and *cis* bend mode in a time interval Δt is:

$$\Delta n_{45} = 4(\sum_{\alpha} q_\alpha dq_\alpha / dt + p_\alpha dp_\alpha / dt)\Delta t$$

$$= 4(\sum q_\alpha \Delta H_{algebraic} / \Delta p_\alpha - p_\alpha \Delta H_{algebraic} / \Delta q_\alpha)\Delta t$$

$$\equiv d_{45}\Delta t \qquad (\alpha = \text{'+', '-'})$$

For a point (q_+, p_+, q_-, p_-) in the phase space, we have the corresponding d_{45}. Suppose that when d_{45} is positive the action is transferred from the *cis* bend to the *trans* bend. The process is reversed when d_{45} is negative. (It does not matter if this definition is in an alternative way) Now we consider those (q_+, p_+, q_-, p_-) for which d_{45} is negative. For each Δt, we calculate $d_{45}\Delta t$. As$|d_{45}\Delta t / N_b| < 0.01$, we consider that action is still in the *trans* mode and set $P(\Delta t) = 1$, otherwise if $|d_{45}\Delta t / N_b| > 0.01$, we consider that action is no longer in the *trans* mode and set $P(\Delta t) = 0$. $P(\Delta t)$ is the survival probability of the *trans* mode as a function of Δt. Of course, we have to average the probability over all points in the phase space. From now on, $P(\Delta t)$ means the averaged survival probability. It is an important dynamical parameter showing the action transfer rate between the *trans* and *cis* modes.

Fig.12.2 shows $P(\Delta t)$ for each energy range denoted in Fig. 12.1.

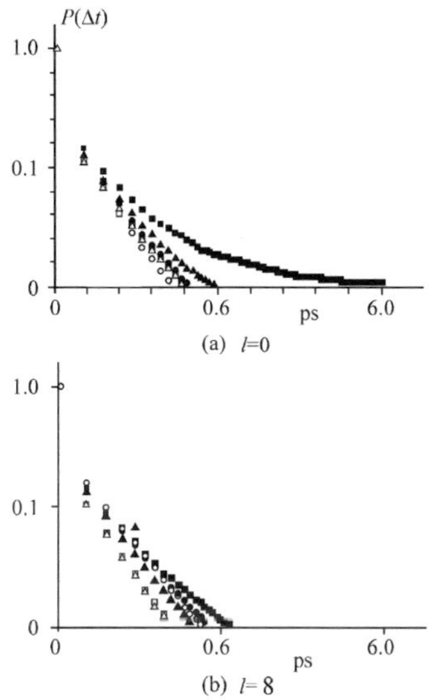

Fig.12.2 $P(\Delta t)$ for each energy range denoted in Fig. 12.1 (a) $l = 0$ (b) $l = 8$

From the Figure, we note that except the very high levels of $N_b = 22$, $l = 0$, $P(\Delta t)$ is not so much dependent on how high a level is and the value of N_b (Careful study reveals that as l is larger, $P(\Delta t)$ tends to be smaller. But for $N_b = 10$, it tends to be larger.). The relaxation time is roughly between 0.45 and 0.6 ps (1 ps = 10^{-12} sec). An exception is that for the highest level of $N_b = 22$, $l = 0$, the relaxation time can extend up to 10 fold, or 6 ps. This implies that faster relaxation is not a necessary consequence of high excitation. This *exceptional* phenomenon is not uncommon in the highly excited system. The core cause is the nonlinearity of inter-mode couplings.

For a system with slower relaxation, its spectrum will be *simpler*, such that the spectral peaks will be narrower, discrete and easier to be recognized. Indeed, in the dispersed fluorescence experiment, as N_b is close to 22, the spectrum due to the relaxation (transition) of the *trans* mode to the *cis* mode, i.e., the fractionation of the *trans* mode on the *cis* mode (this is the overlap of the wavefunction of the *trans* mode on the *cis* mode) is much simpler than in the other cases. Quantum calculation also indicates that as N_b is large, the energy transfer between the *trans* and *cis* modes shows quasiperiodic pattern (Refs. 12.1–3). All these are consistent with our classical results by the coset algorithm. The conclusions are: (1) For high excitation, the classical algorithm can be suitable in some cases. There is no dogma that a strict quantal algorithm has to be followed. (2) The coset algorithm is simple and it grasps the core of the complicated issue. It enables us to obtain the results that can also be obtained by other much more complex algorithms such as the solvation of differential equations. In the following section, we will discuss the modes revealed in the high C–H bend motion, where the advantage of the coset algorithm is evident.

12.7 Modes of C–H bend motion

Previously, it was mentioned that the dynamics of the C–H bend motion of acetylene can be realized by two coupled su(2) algebras, $\{J_{x+}, J_{y+}, J_{z+}\} \otimes \{J_{x-}, J_{y-}, J_{z-}\}$. H_{eff} (or $H_{\text{algebraic}}$) adopts the so-called normal mode picture in which 4 and 5 denote the *trans* and *cis* normal modes, respectively. For the su(2) algebra, we can rotate $\pi/2$ along the J_y axis to form the new $\{J_x', J_y', J_z'\}$ such that $J_x' = -J_z$, $J_y' = J_y$, $J_z' = J_x$. In the new system, 4′, 5′ (those with super prime notation) denote the bend motions of the two C–H bonds, respectively. This is the so-called local mode picture. The advantage of the local mode picture is its intuitiveness.

Now, we have the (q_+', p_+', q_-', p_-') space. Its transformation to (q_+, p_+, q_-, p_-) has been shown in Section 4.3. We can first obtain the solution space (q_+, p_+, q_-, p_-), then shift via transformation to the space in (q_+', p_+', q_-', p_-') to obtain quantities like n_4', n_5', …, etc..

Current experiments are limited to the cases with $l = 0, 2$. Our calculation shows that the results for the cases with zero and nonzero l are not much different. In the following discussion, we will limit ourselves to the case where $l = 0$. In such a case, $P_1 = P_2$ and the system dynamics is symmetric with respect to '+' and '-' motions.

Therefore, only the discussion for the '+' case is necessary. We will stress the relations between Δn_+ ($\equiv (n_{4+} - n_{5+})/P_1$) and φ_+ ($\equiv \tan^{-1}(-p_+/q_+)$ is the angle), Δn_- ($\equiv (n_{4-} - n_{5-})/P_2$) and φ_-, Δn_- and Δn_+, φ_- and φ_+ (The same is true for the local mode picture). One point that needs clarification is that different initial points will lead to different trajectories. Our discussion will be from the viewpoint of the *global* properties. Though trajectories are different, they can show common *global* properties. Therefore, we will only show representative trajectories for discussion. In particular, we will compare the behaviors of the lower and higher levels for various N_b.

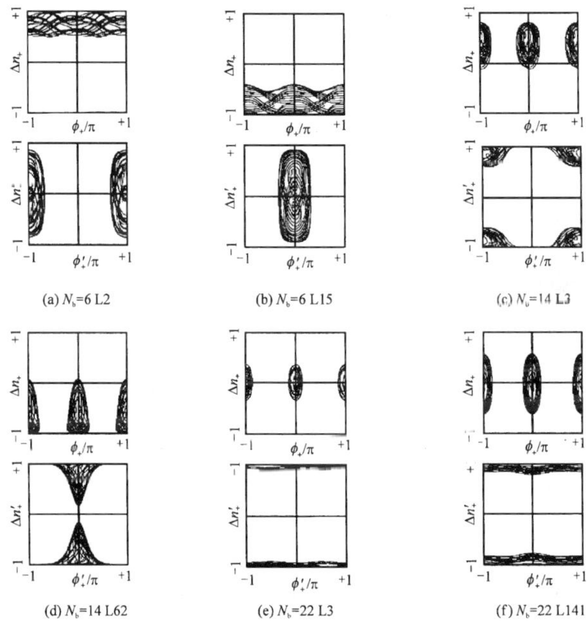

Fig. 12.3 The trajectories in the phase space for different states. See text for details.

(a) $N_b = 6, l = 0$

For $N_b = 6, l = 0$, we have 16 states. They are labeled with L1, L2, …, L16 from the lowest one.

Fig.12.3(a) shows the property of L2. Since $\Delta n_+ \approx P_1$, most action (energy) is stored in the '4' mode, i.e., the *trans* bend. That φ_+' is around $\pm\pi$, implies the phase difference between the two C–H bend motion is π. Hence, it is antisymmetric.(For such an analysis, see Section 4.6) Fig.12.3(b) shows that L15 is a *cis* bend. The two

C–H bends are in phase.

(b) $N_b = 14, l = 0$

For L3 (Fig.12.3(c)), since Δn_+ is close to 1, the mode is close to a *trans* mode. Meanwhile, φ_+' is close to $\pm\pi$. So, these two observations are consistent. However, $\Delta n_+'$ is close to ± 1 and the traces form two discrete regions. This shows that action is concentrated in either the C–H bend and cannot flow freely between the two C–H bends. Note that though this mode is *trans,* it is different from L2 of $N_b = 6$. φ_+ of L2 is extensive between $-\pi$ and π while here φ_+ is limited to 0 and $\pm\pi$. For the high level L62 (Fig.12.3(d)), the mode is *cis* and possesses localized character. Contrary to L15 of $N_b = 6$, φ_+ is limited to 0 and $\pm\pi$.

(c) $N_b = 22, l = 0$

For the low levels like L3, or high levels like L141, (Fig.12.3(e), (f)), φ_+' is extensive and $\Delta n_+'$ is nearly 1, −1. This shows that the two C–H bends are independent with no phase correlation and form a local mode with action concentrated only on either bend. For higher levels like L25, action can flow between the two C–H bends with φ_+' being $\pi/2$. As viewed along the C–C bond, when the angle between the two C–H bends is $\pi/2$, the action transfer is most probable.

In the above discussion, the dynamics of the C–H bend is analyzed from the viewpoint of trajectory. In the following analysis, we will instead discuss the dynamics from a global viewpoint on the phase space..

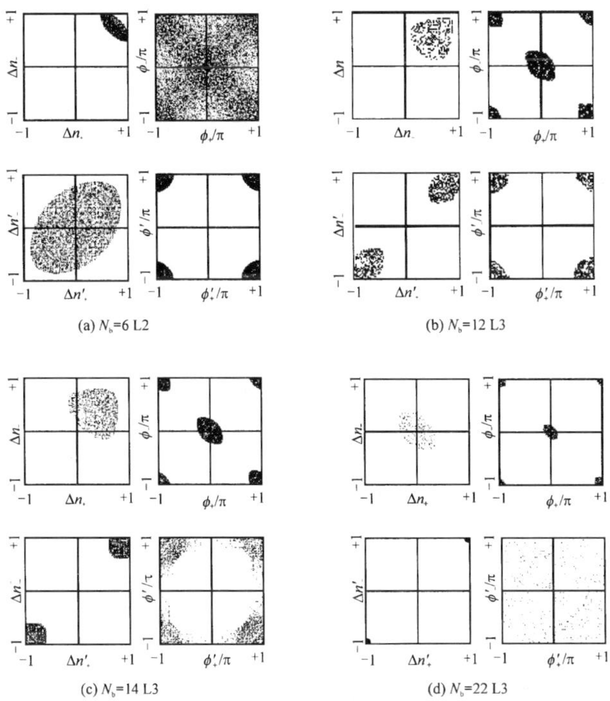

Fig.12.4 The phase space structures for different states. See text for details.

(a) Low levels of $N_b = 6, 12, 14, 22$

We will analyze the mode characters by viewing the relations among: Δn_+, Δn_- ($\Delta n_+'$, $\Delta n_-'$): φ_-, φ_+(φ_-', φ_+'). Fig.12.4(a) shows the result of $N_b = 6$, L = 2. Since $\Delta n_+ = (n_{4+} - n_{5+})$ and $\Delta n_- = (n_{4-} - n_{5-})$ are close to the maximal value $N_b/2$, i.e., both n_{5+} and n_{5-} are close to zero, hence, $n_4 = n_{4+} + n_{4-} = N_b$. The mode is *trans* and normal. This also shows up in that the difference of φ_-', φ_+' is π. The relation between the corresponding φ_-, φ_+ is not definite. Neither is the relation between $\Delta n_-'$ and $\Delta n_+'$. This shows that action can freely flow between the two C–H bends. This is the character of a normal mode. This is consistent with the previous analysis.

The case for $N_b = 12$, L3 is shown in Fig.12.4(b). The mode is *trans*. But, n_4' or n_5' is close to the maximum. The action is concentrated on either the C–H bend and its transfer between the two C–H bends is limited. Since φ_+' and φ_-' are around π, the mode is *trans*. As $N_b = 14$, the mode character of L3 remains the same as shown in Fig. 12.4(c). Now, both φ_+' and φ_-' show dispersive behavior. The phase angle between the two C–H bends is no longer fixed at π. This is the character of a local mode.

As $N_b = 22$, the result for L3 is shown in Fig.12.4(d). n_4 and n_5 are $N_b/2$. Either n_4' or n_5' is N_b. There is no fixed phase relation between φ_-' and φ_+'. These indicate that action is localized on either C–H. Besides, $n_{4+}' = n_{4-}'$ (or $n_{5+}' = n_{5-}'$) shows that the bending motion of C–H is in a plane. This *one hydrogen* bending motion can be visualized as shown in Fig.12.5.

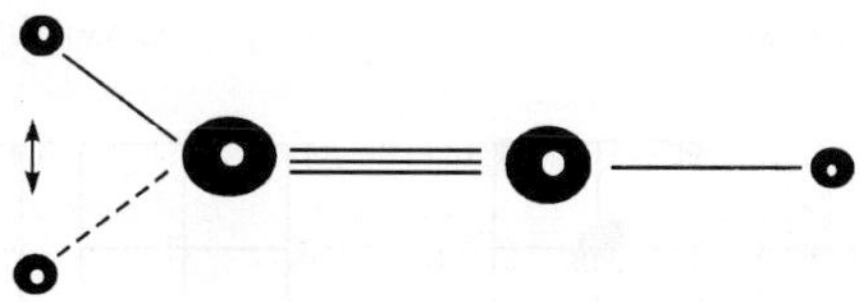

Fig.12.5 One hydrogen bending motion

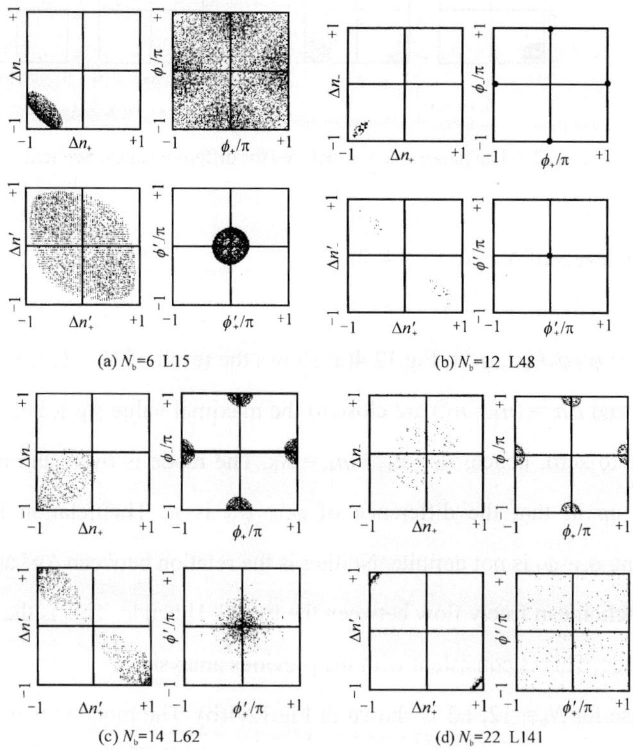

Fig.12.6 The phase space structures for different states. See text for details.

(b) High levels of $N_b = 6, 12, 14, 22$

As $N_b = 6$, the result of L15 is shown in Fig. 12.6(a). From the relation of Δn_- and Δn_+, we know that n_5 is close to N_b and n_4 is very small. This is a *cis* normal mode. That $\varphi_+' \approx \varphi_-' \approx 0$ and no fixed relation between φ_- and φ_+ also supports this assertion. The relation of $\Delta n_+'$ and $\Delta n_-'$ also indicates that action can flow between the two C–H bends.

As $N_b = 12$, the result of L48 is shown in Fig.12.6(b). n_5 is close to N_b and n_4 is almost 0. This is a *cis* mode. However, a difference with the common normal mode is that the phase between φ_-, φ_+ is fixed around $\varphi_+ = \pi$, $\varphi_- = 0$ or $\varphi_+ = 0$, $\varphi_- = \pi$. Besides, n_{4+}' and n_{5-}' are close to $N_b/2$ and correspondingly, n_{4-}' and n_{5+}' are very small; or that both n_{4-}' and n_{5+}' are close to $N_b/2$, while n_{4+}' and n_{5-}' are very small. This shows that the two C–H bends are in opposite rotation with roughly equal action. The energy transfer between the two C–H is not so free since the relation diagram between $\Delta n_+'$ and $\Delta n_-'$ is bisected into two regions. (This can also be seen from the relation of $\Delta n_+'$ and φ_+' or $\Delta n_-'$ and φ_-' that around $\varphi_+' = \varphi_-' = 0$, the diagram (not shown) is separated into two regions) Since $\varphi_+' = \varphi_-' = 0$, as one C–H is in '+' rotation, though the action of another C–H that is in '+' rotation is very small, they are still in phase. This phenomenon is also true for '−' rotation.

As $N_b = 14$, the result of L62 is shown in Fig 12.6(c). The two C–H bends are in opposite rotation. However, the distribution of φ_+', φ_-' is becoming dispersive. The two C–H bends are losing their phase relation in '+' or '-' motion. For L141, $N_b = 22$, Fig.12.6(d) shows that the phase relation no longer exists. Motion is such that the two C–H are in counter rotation, showing evident local character. This particular mode is depicted in Fig. 12.7.

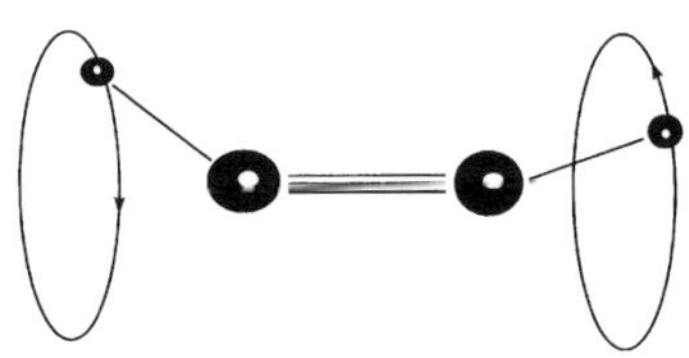

Fig.12.7 Counter rotation of two C-H bend motions

From the aspect of trajectory, there is no apparent difference for L3 and L141 of $N_b = 22$ (See Figs. 12.3(e), (f)). However, from the global aspect of phase space, they are distinctly different. (Compare the relation of $\Delta n_+'$, $\Delta n_-'$ in Figs. 12.4(d), 12.6(d)) The difference lies in that for L3, $n_4' = N_b$ while $n_5' = 0$ (or $n_4' = 0$, $n_5' = N_b$). For L141, $n_{4+}' = n_{5-}' = N_b/2$, $n_{4-}' = n_{5+}' = 0$ (or $n_{4-}' = n_{5+}' = N_b/2$, $n_{4+}' = n_{5-}' = 0$). L3 is one hydrogen bending while L141 is that the two C–H bends are in counter rotation. Apparently, the global approach is superior to the trajectory aspect. This is expected since trajectories are initial condition dependent and the approach based on them is hardly exploited. In summary, the global approach to the phase (solution) space solution is more complete.

We note that the modes as shown in Figs.12.5 and 12.7 have also been reported by a quantal algorithm although without those details about the dynamical configuration and inter-mode action transfer as shown here. (See Refs. 12.3, 12.4)

12.8 Geometric interpretation of vibrational angular momentum

The vibrational angular momentum l caused by the two C–H bends is $l_+' - l_-'$ or $(n_{4+}' + n_{5+}')-(n_{4-}' + n_{5-}')$. It is the difference between the '+ and '−' rotations. Fig.12.8 shows the case of L4 with $N_b = 6$, $l = 2$. The '+' motion possesses larger rotational momentum or angular momentum since for the '−' motion, the phase angle between the two C–H bends can range between $-\pi$ and π, i.e., with no definite phase angle, while for the '+' motion, the phase angle is centred around π (or $-\pi$), i.e., in *trans* configuration with nonzero l. This is depicted in Fig.12.9.

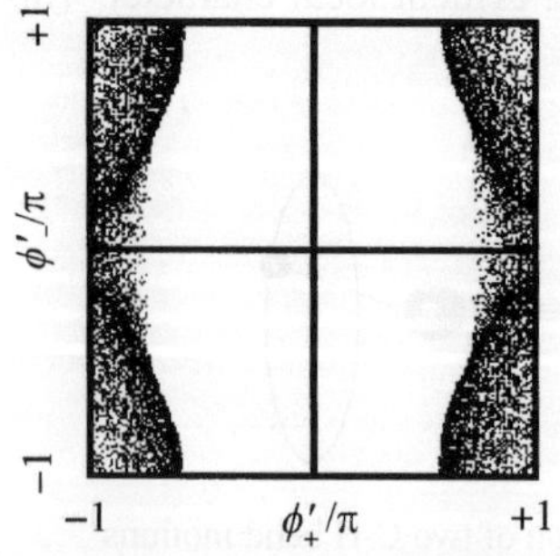

Fig.12.8 The phase relation for the case with nonzero vibrational angular momentum

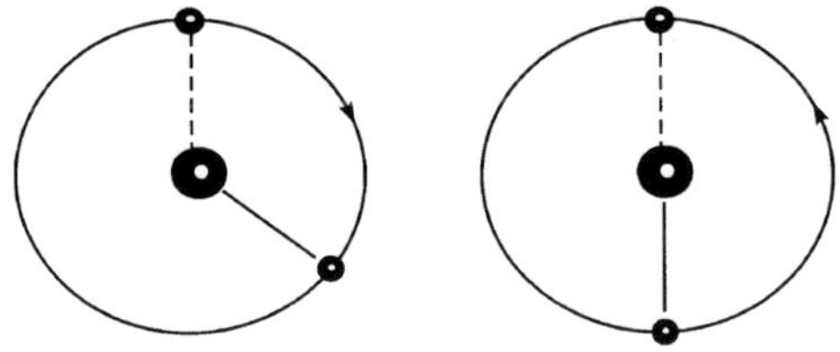

Fig.12.9 For the'-'motion (a), the phase angle between the two C–H bends can range between $-\pi$ and π, i.e., with no definite phase angle, while for the '+'motion (b), the phase angle is centred around π (or $-\pi$).

12.9 Reduced Hamiltonian of C-H bend motion

As mentioned previously, in the formation of the C–H bend motion, vibrational l coupling is not so important as DD-I, DD-II. In general, l is small (only 0, 2) and can be neglected. Therefore, we consider only the DD coupling which can be simplified as $K_{DD}(a_4^{+2}\, a_5^2 + h.c.)$. 4 and 5 stand for the *trans* and *cis* modes. In addition, we consider other modes as a bath for modes 4 and 5. Though modes 4 and 5 are not in resonance with the bath since their frequencies are not compatible, there can be anharmonicity among them. Hence, the *reduced* Hamiltonian is:

$$H_{bath} = \sum_{i=1}^{3} \omega_i (n_i + 1/2) + \sum X_{ii} (n_i + 1/2)^2 + \sum_{i<j} X_{ij} (n_i + 1/2)(n_j + 1/2)$$

$$H_{int} = \sum_{i=1}^{3} X_{i4} (n_i + 1/2)(n_4 + 1/2) + \sum X_{i5} (n_i + 1/2)(n_5 + 1/2)$$

$$H_b = \omega_4 (n_4 + 1/2) + \omega_5 (n_5 + 1/2) + X_{44} (n_4 + 1/2)^2 + X_{55} (n_5 + 1/2)^2$$
$$+ X_{45} (n_4 + 1/2)(n_5 + 1/2) + K_{DD} (a_4^{+2} a_5 + h.c.)$$

The coefficients have been determined in the literature by the fit to the experimental level energies. They are compiled in Refs.12.5,12.6 and are also listed in Table12.2.

Table12.2 The coefficients of the *reduced* Hamiltonian (cm^{-1})

ω_1	3398.74
ω_2	1981.71
ω_3	3316.09
ω_4	609.01
ω_5	729.17
X_{11}	−26.57
X_{22}	−7.39
X_{33}	−27.41
X_{44}	−3.08
X_{55}	−2.34
X_{12}	−12.62
X_{13}	−105.09
X_{14}	−15.58
X_{15}	−10.85
X_{23}	−6.10
X_{24}	−12.48
X_{25}	−1.57
X_{34}	−6.96
X_{35}	−8.69
X_{45}	−2.41
K_{DD}	−2.75

su(2) algebra can be defined by: $J_+ = a_4^+ a_5$, $J_- = a_5^+ a_4$, $J_z = n_4 - n_5$. In the SU(2)/U(1) coset space, we have:

$$n_4 = 2(q^2+p^2), \qquad n_4^2 = 2(q^2+p^2)\{[J-(q^2+p^2)]/J+2(q^2+p^2)\}$$

$$n_5 = 2[J-(q^2+p^2)], \quad n_5^2 = 2J\{2J+(q^2+p^2)/[J-(q^2+p^2)]\}\{[J-(q^2+p^2)]/J\}^2$$

$$n_4 n_5 = 2(2J-1)[J-(q^2+p^2)](q^2+p^2)/J$$

$$J_+^2 + J_-^2 = 4(2-J^{-1})[J-(q^2+p^2)](q^2+p^2)$$

For the reduced Hamiltonian, n_1, n_2, n_3 are given integers and n_4, n_5 are treated as continuous variables. The reduced Hamiltonian is in the so-called normal mode picture. Phase angle $\varphi = \tan^{-1}(-p/q)$ is the phase difference between the *trans* and *cis* modes. We can rotate $\pi/2$ along the J_y axis to obtain the local mode picture in which n_4' and n_5' show the action of the two C–H bends, respectively. $\varphi' = \tan^{-1}(-p'/q')$ is their phase difference. The transformation between (q, p) and (q', p') spaces was described in Chapter 4 in detail.

For an eigenenergy E_i, we can obtain $q = q(p)$ from the equation:

$$E_i = H_{bath}(n_1, n_2, n_3) + H_b(q, p) + H_{int}(n_1, n_2, n_3, q, p)$$

From the solution, the dynamical properties can be easily obtained.

12.10 Mode characters

The mode characteristics of an SU(2) system were described in Chapter 4. Hereby, we summarize the results and point out the exclusive precessional mode in acetylene.

(a) Local mode

In the local mode picture, if φ' is extensive from 0 to 2π, then it is a local mode. In the local mode picture, the mode possesses two separate and equivalent trajectories, one with $n_4' > n_5'$, while the other with $n_4' < n_5'$. They are analogous to the clockwise and counter-clockwise rotations of a pendulum. In the normal mode picture, the phase φ of the local mode is limited to 0 and π. If in a motion cycle, $n_4 > n_5$, then it possesses more of the *trans* character. Otherwise, if $n_4 < n_5$, then the mode has more of the *cis* character. We label them by L_t and L_c, respectively.

(b) Normal mode

The normal mode is different from the local mode. In the local mode picture, φ' is limited to a specific range and the coupling is strong. While in the normal mode picture, φ is extensive in the whole range of $[0, 2\pi]$. If φ' is centred around 0, then the two C–H bends are in phase. This mode is called symmetric and denoted by N_s. Otherwise, if φ' is centred around π ($-\pi$), then the mode is antisymmetric and labeled with N_a. In the normal mode picture, N_s and N_a possess *cis* ($n_5 > n_4$) and *trans* ($n_4 > n_5$) characters, respectively.

(c) Precessional mode

The precessional mode character has φ centred at $\pi/2$ ($-\pi/2$) while φ' is centred in between 0 and π ($-\pi$). If $n_4 > n_5$, the mode is labeled by P_t. If $n_4 < n_5$, then it is labeled with P_c. φ' of P_c is centred in between 0 and $\pi/2$ ($-\pi/2$). φ' of P_t is centred in between $\pi/2$ and π ($-\pi/2$, $-\pi$).

These mode characters are depicted in Fig. 12.10.

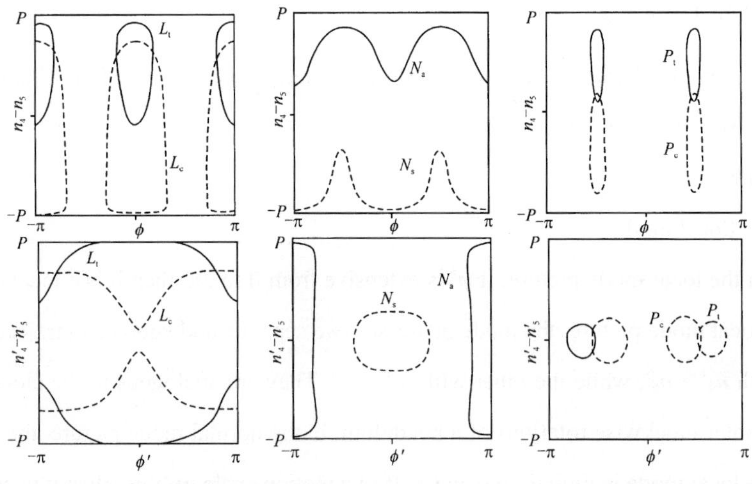

Fig.12.10 The local(a), normal(b) and precessional (c) modes

12.11 Modes of C–H bend motion

For the reduced Hamiltonian of the C–H bend system, n_1, n_2, n_3 are given and $P \equiv n_4 + n_5$ is a conserved quantity ($P = 2J$). For a given P, there are $P/2+1$ levels. In general, for a two-mode system with 1:1 and 2:2 couplings, low levels are of local character while high levels are of normal character. This is true for the C–H bend motion of acetylene though with variations. Table12.3 shows the distribution of mode characters for different P under $(n_1, n_2, n_3) = (0, 0, 0)$. In the Table, the level energy is higher across from L_t, L_c, N_a, N_s, P_t, to P_c. For $P < 12$, all levels are normal. The lower levels are N_a and the higher ones are N_s. The increase of P only results in more levels of N_s character. As P is between 14 and 18, Levels of N_a character disappear and the number of N_s levels remains unchanged. Meanwhile, the number of L_t levels increases. As P reaches 22, both L_t and N_s levels disappear, accompanied by the birth of L_c, N_a and P_c, P_t levels. As P is even larger, there remain only the L_c, N_a and P_t levels and only the number of N_a levels increases; the rest do not increase obviously.

Table12.3 The distribution of mode characters for different P and n_1, n_2, n_3

P	$(n_1, n_2, n_3) = (0, 0, 0)$						$(n_1, n_2, n_3) = (5, 0, 0)$					
	L_t	L_c	N_a	N_s	P_t	P_c	L_t	L_c	N_a	N_s	P_t	P_c
4			1	2					1	2		
6			2	2					2	2		
8			2	3					2	3		
10			2	4					2	4		
12			2	5					2	5		
14	2			6					2	6		
16	3			6			2			7		
18	4			6			3			7		
20	5			4		2	4			7		
22		7		1		4	5			5		2
24		7	3		3		7			3		3
26		7	4		3			8		2		4
28		7	6		2			9	2		4	
30		7	7		2			8	5		3	
	$(n_1, n_2, n_3) = (0, 5, 0)$						$(n_1, n_2, n_3) = (0, 0, 5)$					
4			1	2					1	2		
6			2	2					2	2		
8			2	3					2	3		
10			2	4					2	4		
12			3	4			2			5		
14			3	5			2			6		
16			3	6			3			6		
18			2	8			5			3		2
20	3			8				6		2		3
22	4			8				7	2		3	
24	5			8				6	4		3	
26	6			6		2		6	6		2	
28	7			5		3		6	7		2	
30	9			3		4		6	8		2	

With C–H stretch excitation, like $(n_1, n_2, n_3) = (5, 0, 0)$, $(0, 0, 5)$, the results are shown therein. It seems that the mode distributions for various P follow the same trend. For the C–C excitation, like $(n_1, n_2, n_3) = (0, 5, 0)$, the distribution is also shown therein. As P = 24-30, there are more L_t levels than in the case with C–H stretch excitation. As P is larger, the number of L_t levels also increases by a large amount.

The transition from acetylene to vinylidene is depicted in Fig.12.11.

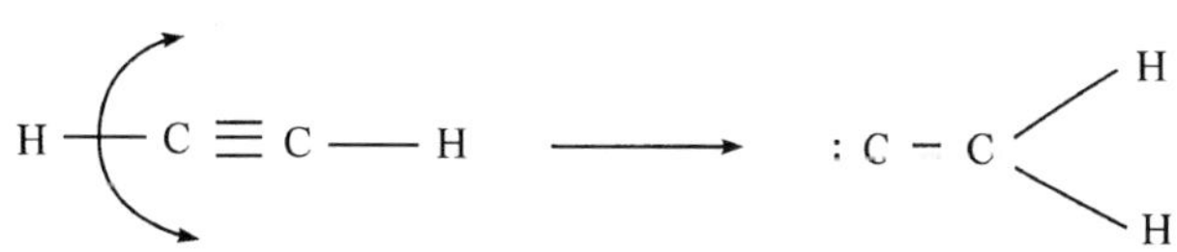

Fig.12.11 The transition from acetylene to vinylidene

The potential barrier by the quantum chemical calculation is about 48 kcal/mole (Ref.

12.7), which is equivalent to about 23–28 quanta on the C–H bend. For such a transition, there are three criteria:

(1) There must be 23–28 quanta on the C–H bend excitation. The modes must be local. Otherwise, the excitation will be dispersed between the two C–H bends. This will lead to action decrease on each C–H bend and the transition to surmount the barrier will be more difficult.

(2) For the local modes, L_t mode is more appropriate since *trans* configuration is prone to the migration of an H atom from one carbon atom to the other without steric hindrance. Therefore, the formation of vinylidene is more probable.

(3) From the viewpoint of experiment, we will have more choices if there are more L_t levels.

From the above consideration, we expect that the accompanying C–C stretch excitation is prone to the transition from acetylene to vinylidene.

There are two points that need attention here. One is that as the frequency of one (or several) mode(s) is far from the rest modes, it is appropriate to consider these modes as a bath with which the mode(s) is anharmonically coupled. This concept could be helpful to the understanding of intramolecular vibrational energy redistribution (IVR). The second is that not much is known about the precessional mode. What its role is in IVR as compared to the local and normal modes is still not much understood at this point in time. This deserves attention.

12.12 su(2) origin of precessional mode

One question is: why is there no precessional mode in the eigenstates of H_{eff} or $H_{\text{algebraic}}$ while in the reduced Hamiltonian (Section 12.9) the precessional mode appears? We will try to understand this from a dynamical viewpoint.

We know that the coupling of two modes α and β can be:

$$a_\alpha^+ a_\beta + a_\alpha^+ a_\alpha^+ a_\beta a_\beta + \ldots + h.c.$$

For this system, we can construct an su(2) algebra $\{J_x, J_y, J_z\}$ as:

$$J_x = (a_\alpha^+ a_\beta + a_\beta^+ a_\alpha)/2,$$

$$J_y = -i(a_\alpha^+ a_\beta - a_\beta^+ a_\alpha)/2,$$

$$J_z = (n_\alpha - n_\beta)/2$$

In terms of the su(2) operators, the interaction is

$$C_1 J_x + C_2 J_x^2 + \ldots$$

For this system, we can rotate $\pi/2$ along the J_y axis so that $J_z' = J_x$, i.e., the quantization is along the x axis. The transformation between $\{J_x, J_z\}$ and $\{J_z', -J_x'\}$ leads to the local and normal mode pictures.

In principle, we can have $C_3 J_y$ (or J_y^2) interaction for which the phase angles of the trajectories, in the normal and local mode pictures, are centred around $\pi/2$. This is the precessional mode as shown in Section 12.10 (c). Fig.12.12 shows the normal, local and precessional modes in the normal and local mode pictures.

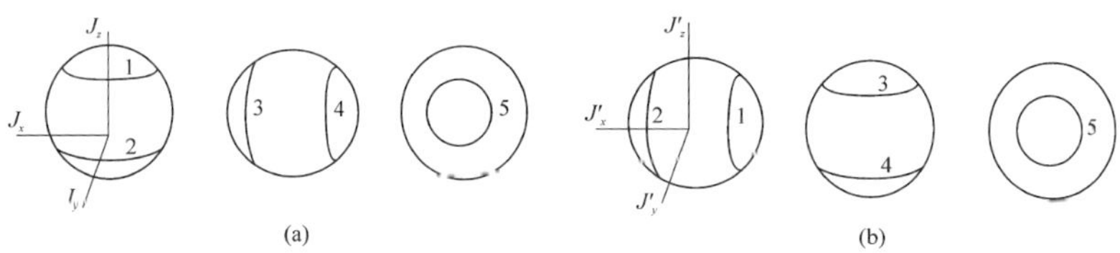

Fig.12.12 (a) Normal mode picture, (b) Local mode picture 1,2 the normal modes、3,4 the local modes and 5 the precessional mode

DD-I, DD-II and vibrational l doubling in H_{eff} can be written in terms of J_x, J_y as:

$$\text{DD-I:} \quad 2\, s_{45}(J_{x+}J_{x-} - J_{y+}J_{y-})$$

$$\text{DD-II:} \quad 1/2\,(r_{45} + 2\, g_{45})\,(J_{x+}^2 + J_{x-}^2 - J_{y+}^2 - J_{y-}^2)$$

$$\text{vibrational } l \text{ doubling:} \quad 2\, r_{45}\,(J_{x+}J_{x-} + J_{y+}J_{y-})$$

Since the coefficient of DD-II is smaller than those of DD-I and vibrational l doubling, $g_{45} \sim - r_{45}$, hence $(r_{45} + 2\, g_{45})/2 \sim - r_{45}/2$. Besides, the high order terms cancel each other out. For simplicity, they are neglected. Since $s_{45} \approx r_{45}$, $J_{y+}J_{y-}$ terms in DD-I and vibrational l doubling cancel each other out. Only the $J_{x+}J_{x-}$ term remains in H_{eff}. Hence, there are only normal and local modes and no precessional mode.

To confirm this assertion, let $r_{45} = - s_{45}$, then H_{eff} only contains terms in $J_{y+}J_{y-}$. Indeed, there is a change from Fig. 12.13(a) with $r_{45} = s_{45}$ ($n_4 + n_5 = 6$, $l = 0$) to (b) with

$r_{45} = -s_{45}$ ($n_4 + n_5 = 22$, $l = 0$). In Fig.(b), trajectories are only around the J_y axis and the precessional mode appears. (Note that Fig. 12.13 is true for both '+' and '−' motion.)

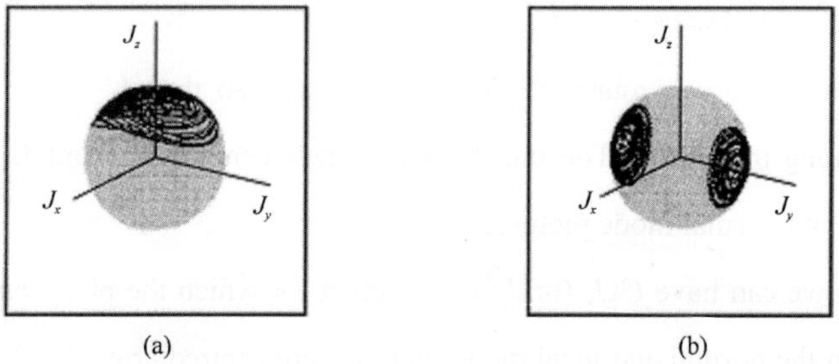

Fig.12.13 (a)The trajectories with $r_{45} = s_{45}$ ($n_4 + n_5 = 6$, $l = 0$) and (b) with $r_{45} = -s_{45}$ ($n_4 + n_5 = 22$, $l = 0$). In (b), trajectories are only around the J_y axis and there appears the precessional mode.

In their study of formaldehyde, Gray and Child proposed the existence of a precessional mode (Ref.12.8). Their Hamiltonian is:

$$H_0(n_+, n_-) + V(n_+, n_-)\cos 4q_-$$

with

$$n_+ = n_s + n_Q, \quad n_- = n_s - n_Q$$

Here, s stands for the coordinate of the dissociation of H_2CO into H_2 and CO. Q is the out-of-plane bend of H_2CO. Since the Hamiltonian is independent of q_+, n_+ is conserved. For a given n_+, the system is an su(2) system. The coupling term $V(n_+, n_-)$ $\cos 4q_-$ was approximated by them as:

$$n_s n_Q \cos 2(q_s - q_Q) \equiv n_s n_Q \cos 2\varphi = n_s n_Q (\cos^2\varphi - \sin^2\varphi)$$

Now, let s, Q be the coordinates 4,5 of the acetylene case, then $n_s + n_Q = 2J$, $n_s - n_Q = 2J_z$, i.e., $n_s = J + J_z$, $n_Q = J - J_z$. Hence, $n_s n_Q = J^2(1-\cos^2\theta) = J^2 \sin^2\theta$ and the coupling term is $n_s n_Q \cos 2(q_s - q_Q) = J^2 \sin^2\theta (\cos^2\varphi - \sin^2\varphi) = J_x^2 - J_y^2$. Clearly, Gray and Child would observe the normal, local and precessional modes in their formaldehyde system.

Rose and Kellman only considered the DD-I mechanism in acetylene, i.e., the $J_{x+}J_{x-} - J_{y+}J_{y-}$ term. Hence, they also observed the precessional mode (Ref.12.9).

In summary, we have reported systematically the theoretical background of the precessional mode. That is: in an su(2) algebraic Hamiltonian, J_x, J_z axes are the limit of the local and normal modes while the J_y term will lead to the precessional mode. Finally, we note that precessional mode was first mentioned by Noid and Marcus in their semiclassical calculation of the Henon–Heiles model with 1:1 coupling.(Ref. 12.10)

References

12.1 M. P. Jacobson, J. P. O'Brien, R. J. Silbey, R. W. Field, *J. Chem. Phys.*, 109 (1998) 121.

12.2 M. P. Jacobson, J. P. O'Brien, R. W. Field, *J. Chem.Phys.*, 109 (1998) 3831.

12.3 M. P. Jacobson, R. J. Silbey, R. W. Field, *J. Chem. Phys.*, 110 (1999) 845.

12.4 M. P. Jacobson, C. Jung, H. S. Taylor, R. W. Field, *J. Chem. Phys.*, 111 (1999) 600.

12.5 D.M. Jonas, S. A. B. Solina, B. Rajaram, R. J. Silbey, R.W. Field, K. Yamanouchi, S. Tsuchiya, *J. Chem. Phys.*, 99 (1993) 7350.

12.6 S. A. B.Solina, J. P. O'Brien, R. W. Field, W. F. Polik, *J. Phys. Chem.*, 100 (1996) 7797.

12.7 W. Chen and C. Yu, *Chem. Phys. Lett.*, 277 (1997) 245.

12.8 S. K. Gray and M. S. Child, *Mol. Phys.*, 53 (1984) 961.

12.9 J. P. Rose and M. E. Kellman, *J. Chem. Phys.*, 105 (1996) 10743.

12.10 D. W. Noid and R. A. Marcus, *J. Chem. Phys.*, 67 (1977) 559.

12.11 G. Wu, *Chem. Phys.*, 252 (2000) 315.

12.12 G. Wu, *Chem. Phys.*, 269 (2001) 93.

12.13 J. Yu and G. Wu, *J. Chem. Phys.*, 113 (2000) 647.

Chapter 13

Lyapunov exponent and nonergodicity of C–H bend motion in acetylene

13.1 Lyapunov exponent

One characteristic of a chaotic system is that its trajectories are initial-condition dependent. A Lyapunov exponent is a parameter for describing this property.

Suppose that there are two initial points with a deviation $\Delta X(0)$. At time t, the separation between the two evolving trajectories is $\Delta X(t)$. Suppose $\Delta X(t)$ is of the form:

$$\Delta X(t) \sim e^{\lambda t} \Delta X(0)$$

Parameter λ is the Lyapunov exponent. $\lambda < 0$ shows that the two trajectories are converging toward each other. $\lambda = 0$ shows that they are parallel while $\lambda > 0$ shows that they are divergent. $\lambda > 0$ is the condition for a system to be chaotic. λ is defined in a long duration as:

$$\lambda = \lim_{t \to \infty} \ln[\Delta X(t) / \Delta X(0)] / t$$

For a one-dimensional map M, we have

$$X_{n+1} = M(X_n)$$

Suppose in the neighborhood of X_n, there is another point with a deviation ΔX_n, then we have:

$$\Delta X_{n+1} = \left. \frac{dM}{dX_n} \right|_{X_n} \Delta X_n$$

For convenience, write $\left. \frac{dM}{dX_n} \right|_{X_n}$ as $M'(X_n)$, then

$$\Delta X_{n+1} = M'(X_n) \Delta X_n$$

After the n-th map, ΔX_n is

$$\Delta X_n = M'(X_{n-1})\ldots M'(X_0)\Delta X_0 \equiv DM^n(X_0)\Delta X_0$$

then

$$\lambda = \lim_{n\to\infty}\frac{1}{n}\ln(\Delta X_n / \Delta X_0)$$
$$= \lim_{n\to\infty}\frac{1}{n}\ln DM^n(X_0)$$

Consider an initial point X_0 in an N dimensional system with a deviation ΔX_0 along direction u_0 ($|u_0|=1$), then λ will be dependent on X_0 and u_0. And we have

$$\lambda(X_0,u_0) = \lim_{n\to\infty}\frac{1}{n}\ln(\Delta X_n / \Delta X_0)$$
$$= \lim_{n\to\infty}\frac{1}{n}\ln|DM^n(X_0)\cdot u_0|$$

For an initial u_0, after each map, its direction and magnitude will vary. Hence, the deviation in the above formula is along the direction $u_i/|u_i|$ $(i=0,1,\ldots,n-1)$ after i maps starting from u_0. In fact, we have $DM^i(X_0)\cdot u_0 = u_i$ or $M'(X_i)\cdot u_i = u_{i+1}$. Note that DM^n is a matrix which is along the tangent space at X_0 and u_0 is a vector.

As n is large enough, we have

$$\lambda(X_0,u_0) \cong \lambda_n(X_0,u_0) \equiv \frac{1}{n}\ln|DM^n(X_0)\cdot u_0|$$
$$= \frac{1}{2n}\ln[u_0^{\ T}\cdot H_n(X_0)\cdot u_0]$$

Here, $H_n(X_0)=[DM^n(X_0)]^T\, DM^n(X_0)$. Superscript T denotes the transpose of a matrix or a vector. Since $H_n(X_0)$ is real and nonnegative, its eigenvalues are real and nonnegative and its eigenvectors e_{nj} ($j=1,\ldots,N$) can be chosen as real and orthogonal. Denote the eigenvalues of $H_n(X_0)$ as h_{nj} and let $\lambda_{nj} \equiv (2n)^{-1}\ln h_{nj}$. We arrange them in such a sequence:

$$\lambda_{n1}(X_0) \geq \lambda_{n2}(X_0) \geq \ldots \geq \lambda_{nN}(X_0)$$

Therefore, as n is approaching infinity, the Lyapunov exponents $\lambda_j(X_0)$ will also show the sequence:

$$\lambda_1(X_0) \geq \lambda_2(X_0) \geq \ldots \geq \lambda_N(X_0)$$

In general, u_0 is

$$u_0 = \sum_{j=1}^{N} a_{nj} e_{nj}$$

Then,

$$u_0^T H_n u_0 = \sum a_{nj}^2 h_{nj} = \sum a_{nj}^2 \exp[2n\lambda_{nj}(X_0)]$$

$$= a_{n1}^2 \exp[2n\lambda_{n1}(X_0)]$$

The last step holds since as $n \to \infty$, the quantity will be determined by the largest λ_{n1}.

Hence,

$$\lim_{n\to\infty} \frac{1}{2n} \ln[u_0^T H_n(X_0) u_0] = \lambda_1(X_0)$$

Therefore, for an arbitrarily chosen u_0, we will obtain $\lambda_1(X_0)$, i.e., the maximal Lyapunov exponent. For elucidating $\lambda_2(X_0)$, we have to limit u_0 in a space orthogonal to $e_{\infty 1}$. Then by an analogous algorithm for $\lambda_1(X_0)$, we can obtain $\lambda_2(X_0)$. Limiting u_0 again in a space orthogonal to both $e_{\infty 1}$ and $e_{\infty 2}$, we obtain $\lambda_3(X_0)$. Following this procedure, all the Lyapunov exponents can be obtained.

For a periodic trajectory of period p, under map M, we have

$$M\colon X_0^* \to X_1^* \to \cdots \to X_p^* = X_0^*$$

Suppose h_i is the eigenvalue of DM^p (at any X_j^*), then the Lyapunov exponent is

$$\lambda_i = \ln|h_i|/p \quad (i = 1,\ldots,N)$$

When using a computer to calculate a Lyapunov exponent, no matter how small the initial ΔX_0 is chosen, if $\lambda_1(X_0) > 0$, then very soon ΔX_n will be so large that it exceeds the memory. This can be solved by the following method. For example in calculating λ_1, suppose initial ΔX_0 is along a certain direction and the mapping interval is τ. At time τ (initial time is 0), $\Delta X_1 = \alpha_1$. At X_1, along the deviation direction, we divide ΔX_1 by α_1. In other words, we do the normalization before the next mapping. Then, we have $\Delta X_2 = \alpha_2$. At X_2, along the deviation direction, divide ΔX_2 by α_2 and use it as the deviation for the next mapping. After n mappings, the total deviation is $\alpha_n \alpha_{n-1} \cdots \alpha_1$, i.e.,

$$\lambda_{n1}(X_0) = \frac{1}{n\tau} \ln \alpha_n \alpha_{n-1} \cdots \alpha_1$$

$$= \frac{1}{n\tau} \sum_{r=1}^{n} \ln \alpha_r$$

As n is large enough, we have:

$$\lambda_1(X_0) \cong \lambda_{n1}(X_0)$$

The algorithm for calculating a Lyapunov exponent is not unique. Suppose the initial deviation is d_0, after one map, the deviation is d_1. Between the two points, $X_1(\tau)$ and $X'(\tau)$, choose a point $X''(\tau)$ which deviates from $X_1(\tau)$ by d_0. Then we do the mapping on these two points to obtain the new deviation d_2. Following this procedure, we obtain a series of d_1, d_2, ... d_n and

$$\lambda_{n1}(X_0) = \frac{1}{n\tau} \sum_{r=1}^{n} \ln(d_r / d_0)$$

This is a generic formula. As d_0=1, we retrieve the previous formula.

This algorithm can be generalized to calculate $\lambda_2(X_0)$. Suppose the initial deviation is limited to the space by $e_{\infty 1}$, $e_{\infty 2}$ and at time τ, the area enclosed by the deviated vectors is $\alpha_1^{(2)}$. Normalize the area (by the Gram-Schmidt orthonormalization method) to a square with side 1. Then do the mapping to obtain the area $\alpha_2^{(2)}$ formed by the deviated vectors. Repeating the normalization and mapping, we have $\alpha_3^{(2)}$, $\alpha_4^{(2)}$, ... Then

$$\lambda_{n1}(X_0)+\lambda_{n2}(X_0)=\frac{1}{n\tau}\sum_{r=1}^{n}\ln\alpha_r^{(2)}$$

Comparing with $\lambda_{n1}(X_0)$, we have:

$$\lambda_{n2}(X_0)=\frac{1}{n\tau}\left(\sum\ln\alpha_r^{(2)}-\sum\ln\alpha_r\right)$$

$$=\frac{1}{n\tau}\sum\ln(\alpha_r^{(2)}/\alpha_r)$$

By this formula, the extension to obtain every $\lambda_i(X_0)$ is easy.

In the following, we demonstrate the algorithm to calculate $\lambda_i(X_0)$ for a set of differential equations of *N* variables.

The set of differential equations is:

$$\dot{x}_i = f_i(x_1,\ldots x_N),\ i=1,\ldots,N$$

For Hamilton's equations of motion, x_i is q_α or p_α. f_i is $\partial H/\partial q_\alpha$ or $\partial H/\partial p_\alpha$.

Choose a point in this *N* dimensional space: $X_0=(x_1(0),\ldots x_N(0))$. By the equations, we integrate out the trajectory: X_1, X_2, ..., X_n, ...

$$x_i(n+1)=x_i(n)+f_i(X_n)\tau$$

n = 0,1,2,..., shows the time of integration. $x_i(n)$ is the component of X_n. τ is the time interval for integration. It should be properly small.

At X_0, choose an orthonormal *N* dimensional basis set:

$e_1(0)$= (1, 0, 0, ...),

$e_2(0)$= (0, 1, 0, ...),

...

$$e_N(0) = (0,\ 0,\ 0,\ \ldots 1),$$

The end points of these $e_j(0)$ can be taken as deviations: $\delta x_{ji}(0) = \delta_{ji}$. At time τ, from the equations by which the deviations follow, we have:

$$\frac{d}{dt}(\delta x_{ji}) = \sum_k \frac{\partial f_i}{\partial x_k} \delta x_{jk}$$

We then obtain:

$$\begin{aligned} \delta x_{ji}(1) &= \delta x_{ji}(0) + \sum_k \left.\frac{\partial f_i}{\partial x_k}\right|_{X(0)} \delta x_{jk}(0) \\ &= \delta_{ji} + \sum_k \left.\frac{\partial f_i}{\partial x_k}\right|_{X(0)} \delta_{jk} \\ &= \delta_{ji} + \left.\frac{\partial f_i}{\partial x_j}\right|_{X(0)} \end{aligned}$$

This means that the original orthonormal bases $e_j(0)$ are no longer orthonormal after evolution for a duration of τ. Denote them as :

$$v_1(1),\ v_2(1),\ \ldots v_N(1)$$

By the Gram-Schmidt method, we have the new orthonormal bases $e_1(1), \ldots e_N(1)$ as

$$e_1(1) \equiv v_1(1) / |v_1(1)|$$

$$e_2(1) \equiv [v_2(1) - (v_2(1), e_1(1))e_1(1)] / |v_2(1) - (v_2(1), e_1(1))e_1(1)|$$

$$\ldots$$

$$e_N(1) \equiv [v_N(1) - \sum_{i=1}^{N-1}(v_N(1), e_i(1))e_i(1)] / |v_N(1) - \sum_{i=1}^{N-1}(v_N(1), e_i(1))e_i(1)|$$

Define

$$M_1(1) \equiv |v_1(1)|$$

$$M_2(1) \equiv |v_2(1) - (v_2(1), e_1(1))e_1(1)|$$

$$\ldots$$

$$M_N(1) \equiv |v_N(1) - \sum_{i=1}^{N-1}(v_N(1), e_i(1))e_i(1)|$$

Then, take the end points of $e_j(1)$ at X_1 as the deviations, and repeat the procedure to obtain $v_j(2)$, $e_j(2)$, $M_j(2)$, $v_j(3)$, $e_j(3)$, $M_j(3)$,... (j =1, ..., N). Finally, Lyapunov exponent $\lambda_j(X_0)$ is:

$$\lambda_j(X_0) = \lim_{n \to \infty} \sum_{r=1}^{n} \ln M_j(r) / n\tau$$

13.2 Important concepts of a Lyapunov exponent

(1) Though a Lyapunov exponent is initial-point dependent, in general, it is in fact a global property. This is shown in its definition: the long time behavior (limit $n \to \infty$) is required.

(2) All the points on a trajectory possess the same Lyapunov exponent. For the same reason, if two points share the same Lyapunov exponent, then they are very probably on the same trajectory. In a space, if all the points share the same Lyapunov exponent, then starting from any point, the trajectory will run over all the points in space. This is the property of ergodicity. Obviously, a space possessing various Lyapunov exponents cannot be ergodic. Hence, the space is divided into regions, and each possesses a common Lyapunov exponent.

(3) When a Lyapunov exponent is larger than zero, the trajectory is initial-point dependent and chaotic. For an unstable periodic trajectory, its Lyapunov exponent will be larger than zero. Most often, the unstable periodic trajectory is embedded in the sea of chaotic trajectories.

(4) If all the points in a space share identical M', then the sum of Lyapunov exponents at any point is equal to the determinant $|M'|$ which is a constant. This is a sum rule.

(5) As time elapses, although the phase space of a Hamiltonian system will change its shape, its volume will remain unchanged. The Lyapunov exponents of any point in a Hamiltonian phase space are paired as (λ_i, -λ_i). For example, for a five dimensional space, the five Lyapunov exponents are paired as (λ_1, λ_2, 0, -λ_2, -λ_1).

Hence, there are only two independent exponents: λ_1 and λ_2. This will greatly reduce our labor in the calculation of Lyapunov exponents.

(6) For two initially indistinguishable trajectories, if the Lyapunov exponent is larger than zero, their separation will be obvious after a certain time. This means that as time elapses, the information is being created. In other words, the Lyapunov exponent is related to the system information. Kolmogorov and Sinai have proposed the concept of entropy, h_{ks}. h_{ks}, the parameter characterizing dissipation γ (dissipation ~ $e^{-\gamma t}$) and positive Lyapunov exponents λ_i^+ possess the following relation:

$$h_{ks} + \gamma = \sum_i \lambda_i^+$$

For a Hamiltonian system without dissipation, γ =0. Then,

$$h_{ks} = \sum_i \lambda_i^+$$

h_{ks} characterizes the acquired information in each map. It is the sum of all the positive Lyapunov exponents. In other words, entropy (in thermodynamics) is nothing but the sum of positive Lyapunov exponents.

(7) Suppose K is the largest integer such that

$$\sum_{j=1}^{K} \lambda_j \geq 0$$

and we define Lyapunov dimension D_L as

$$D_L = K + \frac{1}{|\lambda_{K+1}|} \sum_{j=1}^{K} \lambda_j$$

Kaplan and Yorke have proposed the conjecture that

$$D_L = D_1$$

D_1 is the information dimension (See Section 11.3). Hence, the Lyapunov exponent is an important parameter for a chaotic system. It is a dynamical concept while information dimension is a geometric concept. This conjecture bridges the relation between the dynamical and geometric aspects of a chaotic system.

(8) It is an easy task to calculate a Lyapunov exponent using a software algorithm. The most difficulty is given by the slow convergence of λ_{nj} as n increases. As λ_{nj} varies with n in the form $\alpha n^{-\beta}$ $(\alpha, \beta > 0)$ or that $\ln \lambda_{nj}$ and $\ln n$ follows a linear relation of $\ln \alpha - \beta \ln n$, then as $n \to \infty$, λ_{nj} will approach to zero. That is, the Lyapunov exponent is zero.

(9) In Section15.9, we will point out that in the phase space where the quantum numbers are approximately conserved, the corresponding Lyapunov exponents are also less. In Chapter 17, we will discuss the concept of quantizing a nonintegrable, chaotic system by minimizing the Lyapunov exponent. All these show that the Lyapunov exponent is an essential parameter for a chaotic system.

13.3 Nonergodicity of C–H bend motion

The coset represented Hamiltonian of C–H bend motion of acetylene, $H_{eff}(q_+, p_+, q_-, p_-)$ was introduced in Chapter 12. For each eigenenergy E, we can obtain the solution space from $H_{eff}(q_+, p_+, q_-, p_-) = E$. From a point in the solution space, by Hamilton's equations of motion, a trajectory can be followed. The trajectory is three dimensional due to energy conservation. For convenience, the surface of section $(q_+, p_+, q_- = 0, p_- < 0)$ can be chosen for viewing the trajectory, i.e., as $q_- = 0$, and $p_- < 0$, the corresponding (q_+, p_+) values are recorded. (For convenience, q_+, p_+ are normalized by $\sqrt{N_b / 2}$ to the range of [-1, 1]. We will only consider the case of $l = 0$). The Lyapunov exponent for each trajectory is also calculable. For this system, we only have to calculate the largest Lyapunov exponent, $\lambda_{\max}$ because the three exponents are paired as $\lambda_{\max}$, 0, $-\lambda_{\max}$. ($\lambda_{\max}$ is just λ_1 shown in the last section)

Our integration step of Hamilton's equations is 3.33 fs, the number of integrations is $2*10^5$, equivalent to 666 ps. This time is much longer than the vibrational relaxation time which is less than 1–10 ps. (The unit of the coefficients of Hamilton's equations is cm^{-1}. Time interval $\Delta t = 1$ is equivalent to $33.3/2\pi$ ps.) In calculating λ_{max}, the initial deviation, d_0, is 10^{-6}. Shown in the Appendices are the Adams-Bashforth-Moulton method for integration and the numerical techniques for recording on the surface of a section.

By numerical analysis, it is hard to exploit all the cases. Hence, the conclusions are hardly perfect. For compensation, initial points are chosen randomly and what we are concerned with is to make the conclusions as generic as possible from *finite* cases. In this respect, numerical analysis is still beneficial.

Numerical conclusions show that the C–H bend motion can be:

(1) Regular motion

For $N_b = 6$ (N_b is the total action or quantum number of bend motion), there are 16 levels and these are labeled from the lowest one with $L\#$ (# is $1, \cdots, (N_b/2+1)^2$). All these levels possess zero λ_{max}, showing that their trajectories are regular. Fig.13.1 shows the surface of a section of a representative trajectory of L9 ($3997.1cm^{-1}$) and its relation between $\log\lambda_{n1}$ and $\log n$. Obviously, its form is $\alpha n^{-\beta}$ and $\lambda_{max} = 0$.

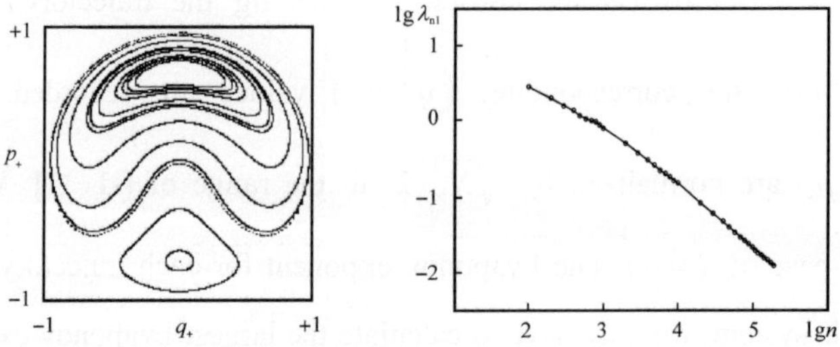

Fig.13.1 The surface of a section of a representative trajectory of L9 ($3997.1cm^{-1}$) and its relation between $\log\lambda_{n1}$ and $\log n$. Obviously, its form is $\alpha n^{-\beta}$ and $\lambda_{max} = 0$.

(2) Irregular motion

For each N_b between 8 and 22, except those higher and lower levels for which trajectories are regular, the phase spaces are full with trajectories with various λ_{max}. For example, for N_b=10, L17 (6602.2 cm^{-1}), λ_{max} ranges from 0.18 ps^{-1} to 0.36 ps^{-1}. Its surface of section is shown in Fig. 13.2 with complex structure.

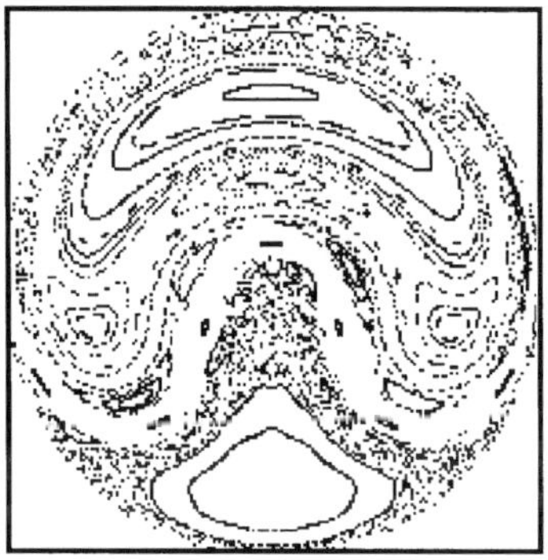

Fig.13.2 The complex surface of section of N_b=10, L17 (6602.2 cm^{-1})

Fig.13.3 shows the surfaces of a section of two trajectories and their relation of $\log\lambda_{n1}$ and $\log n$ for N_b=12, L21 (7900.7cm^{-1}). Their λ_{max} are 0.86 ps^{-1} and 0.32 ps^{-1}, respectively.

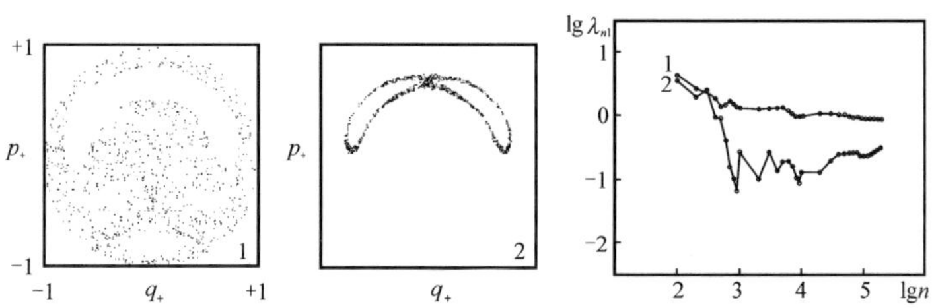

Fig.13.3 The surfaces of a section of two trajectories and their relation of $\log\lambda_{n1}$ and $\log n$ for N_b=12, L21 (7900.7cm^{-1})

Since there is a distribution of λ_{max}, the motion is not ergodic. For each N_b,

$\lambda_{\max}$ increases in general from lower to higher levels and then decreases. Also, as N_b is larger, $\lambda_{\max}$ is larger. For a quantitative concept, for a given N_b, we record the maximal $\lambda_{\max}$ of all its levels and denote it as Λ (in ps^{-1}). We have (Λ, N_b) = (0.06, 8), (0.36, 10), (0.93, 12), (1.56, 14), (2.16, 18), (2.43, 22). Λ is larger as N_b increases.

Though the general cases are nonergodic, there are cases of ergodicity. For example, for N_b = 14, L8 (9070.9cm^{-1}), all the trajectories in the phase space share the same $\lambda_{\max}$. Fig.13.4 shows the traces of three trajectories and their plots of $\log\lambda_{n1}$ against $\log n$. Indeed, they share the same $\lambda_{\max}$.

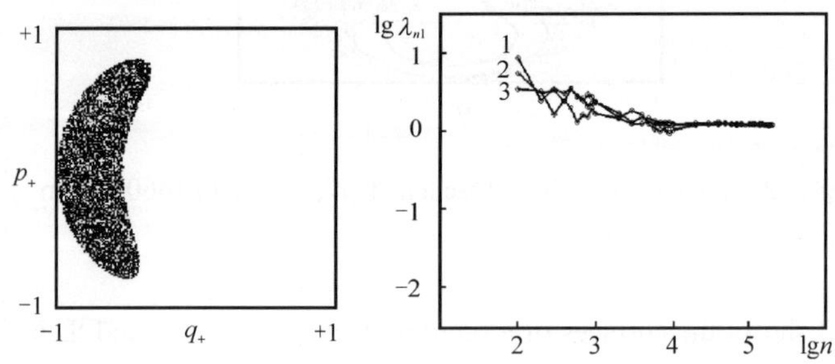

Fig.13.4 The traces and the plots of $\log\lambda_{n1}$ against $\log n$ for three different trajectories that share the same $\lambda_{\max}$

(3) Transition between different degrees of chaoticity

An interesting case is the transition between different degrees of chaoticity for a trajectory as time elapses. Fig.13.5 shows two cases:

Fig. 13.5 (a) is the case with N_b=14, L46 (9435.7cm^{-1}). At 300 ps. $\lambda_{\max}$ jumps from 0.43 ps^{-1} to 0.93 ps^{-1}. The degree of chaoticity enhances. Fig. 13.5 (b) is the case with N_b=18, L6 (11576.7cm^{-1}). At 167 ps, $\lambda_{\max}$ drops from 0.72 ps^{-1} to 0.25 ps^{-1}. A and B in the Figure show these two stages. The degree of chaoticity varies. The corresponding surfaces of a section with different patterns are also shown in the Figure. (Smaller λ_{max} corresponds to smaller region in the surface of section). In the numerical calculation, the energy deviation is less than 0.1cm^{-1}. Hence, different λ_{max} and the different patterns on the surface of the section in the A and B stages are not numerical artifacts. This phenomenon may not be important in reality if we recognize that the time scale of vibrational relaxation is only around 1 ps. However, its physics are noticeable.

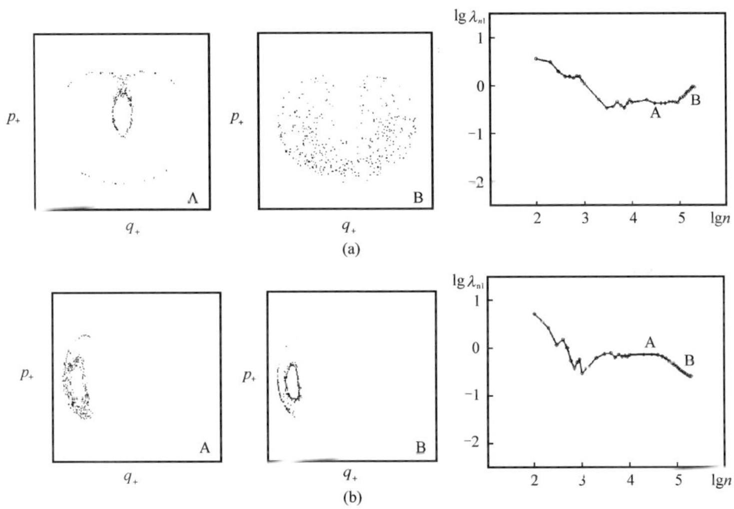

Fig.13.5 Two cases of the transition between different degrees of chaoticity. See text for details.

(4) Periodic trajectories

Up to the time range of 666 ps, we observed the stable and unstable periodic trajectories.

(a) Stable periodic trajectories in $N_b = 12$, L47 (8373.5 cm^{-1}) and $N_b = 6$, L9 (3997.1 cm^{-1}). Viewed from the surface of section, the periodic trajectories are immersed in the quasi-periodic trajectories. Fig.13.6 shows the case of L47 for which the periodic trajectory is the central point. Its λ_{max} is 0.

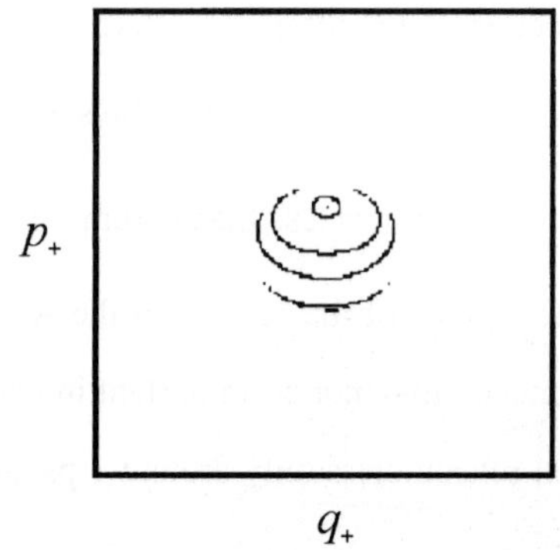

Fig.13.6 The case of L47, $N_b = 12$ for which the periodic trajectory is the central point.

Fig.13.7 shows the relation of $n'_{4-} - n'_{5-} \equiv \Delta n'_-$ and $n'_{4+} - n'_{5+} \equiv \Delta n'_+$ for L47. (4, 5 denote the two C–H bends. + and – denote the right and left rotations. The superscript ' ' ' denotes the local mode picture. See Chapter 12.) Since the slope is -1, $n'_{4+} - n'_{5+} = -(n'_{4-} - n'_{5-})$ or $n'_4 = n'_5$, i.e., the actions on the two C–H bends are the same. For the case of L9 (not shown), the slope is 1 and $l'_4 = l'_5$, i.e., the vibrational angular momenta of the two C–H bends are always the same.

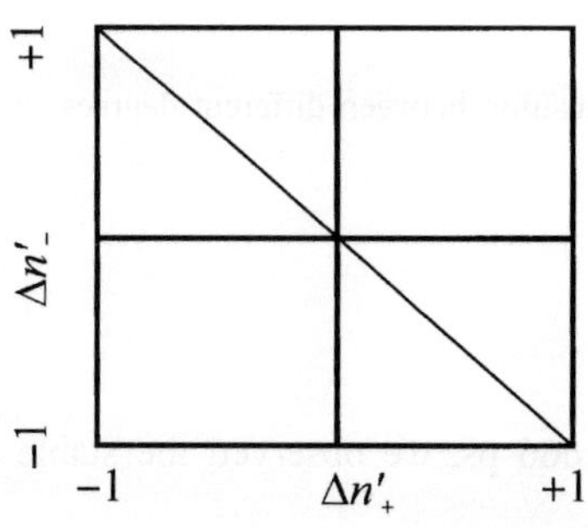

Fig.13.7 The relation of $n'_{4-} - n'_{5-} \equiv \Delta n'_-$ and $n'_{4+} - n'_{5+} \equiv \Delta n'_+$ for L47

(b) Unstable periodic trajectory in $N_b = 14$, L46 (9435.7 cm^{-1}). Fig.13.8 shows the surface of section and the unstable periodic trajectory, marked with $\times$, is among the chaotic sea for which λ_{max} =1.18 ps^{-1} and $n_4' = n_5'$. That is, the actions on the two C–H bends are the same.

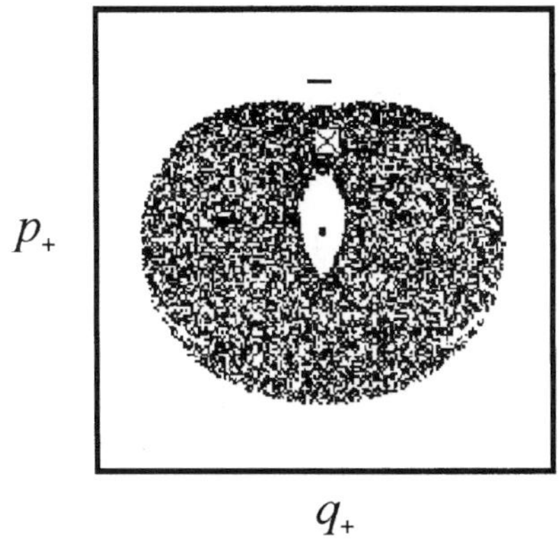

Fig.13.8 The surface of the section and the unstable periodic trajectory, marked with ×, among the chaotic sea N_b = 14, L46 (9435.7 cm^{-1}).

References

13. 1 E. Ott, *Chaos in Dynamical Systems*, Cambridge University Press, Cambridge, 1994.

13. 2 J. R. Dorfman, *An Introduction to Chaos in Nonequilibrium Statistical Mechanisms*, Cambridge University Press, Cambridge, 1999.

13. 3 G. Benettin, L. Galgani, A. Giorgilli, J.-M. Strelcyn, *Meccanica* 15(1980)9.

13. 4 G. Benettin, L. Galgani, J.-M. Strelcyn, *Phys. Rev. A* 14(1976) 2338.

13. 5 J. Yu and G. Wu, *Chem. Phys. Lett.* 343(2001)375.

Appendices:

A. The integration of Hamilton's equations

For the numerical integration of Hamilton's equations, we need to make sure that energy is conserved. Adams-Bashforth-Moulton method was adopted in our work. This method quotes the Runge-Kutta algorithm which is simpler and efficient enough if very high accuracy is not anticipated.

1. Runge-Kutta algorithm

➡ Suppose the differential equations are:

$$
\begin{aligned}
y_0' &= f_0(t, y_0, y_1, \ldots, y_{n-1}) \\
y_1' &= f_1(t, y_0, y_1, \ldots, y_{n-1}) \\
&\vdots \\
y_{n-1}' &= f_{n-1}(t, y_0, y_1, \ldots, y_{n-1})
\end{aligned}
$$

(initial values at time t_0: $y_0(t_0) = y_{00}$, $y_1(t_0) = y_{10}$, …, $y_{n-1}(t_0) = y_{n-1,0}$)

➡ Suppose at time t_j: $y_0(t_j) = y_{0j}$, $y_1(t_j) = y_{1j}$, … $y_{n-1}(t_j) = y_{n-1,j}$:

$$\Rightarrow k_{0i} = hf_i(t_j, y_{0j}, y_{1j}, \ldots, y_{n-1,j})$$

$$\text{from } k_{0i} \Rightarrow k_{1i} = hf_i(t_j + \frac{h}{2}, y_{0j} + \frac{h}{2}k_{00}, \ldots, y_{n-1,j} + \frac{h}{2}k_{0,n-1})$$

$$\text{from } k_{1i} \Rightarrow k_{2i} = hf_i(t_j + \frac{h}{2}, y_{0j} + \frac{h}{2}k_{10}, \ldots, y_{n-1,j} + \frac{h}{2}k_{1,n-1})$$

$$\text{from } k_{2i} \Rightarrow k_{3i} = hf_i(t_j + h, y_{0j} + hk_{20}, \ldots, y_{n-1,j} + hk_{2,n-1})$$

(with $i = 0, 1, \ldots, n\text{-}1$)

➡ from k_{0i}, k_{1i}, k_{2i}, k_{3i}, we obtain the values at $t_{j+1} = t_j + h$:

$$y_{i,j+1} = y_{ij} + \frac{1}{6}(k_{0i} + 2k_{1i} + 2k_{2i} + k_{3i}) \qquad i = 0, 1, \ldots, n\text{-}1$$

➡ from $y_{i,j+1}$ $(i = 0, 1, \ldots, n-1)$ at t_{j+1}, repeat the above procedure:

↖

....

For our equations, $n= 4$, y_i ($i = 0, 1, 2, 3$) are $q_+(t)$, $p_+(t)$, $q_-(t)$, $p_-(t)$; $f_i(t, y_0, y_1, \ldots, y_{n-1})$ are $\partial H_{\text{eff}} / \partial p_+$, $-\partial H_{\text{eff}} / \partial q_+$, $\partial H_{\text{eff}} / \partial p_-$, $-\partial H_{\text{eff}} / \partial q_-$, for which time is not explicit. The integration step is h=10^{-4}.

In applying the Runge-Kutta algorithm, variable steps are adopted:

➡ By h, from $y_i(t_j)$ to obtain $y_i(t_j+h)^{(h)}$

➡ By $h/2$ for two steps, from $y_i(t_j)$ to obtain $y_i(t_j+h)^{(h/2)}$

? If : $\max_{(i = 0, 1, \ldots, n-1)} | y_i(t_j+h)^{(h/2)} - y_i(t_j+h)^{(h)} | < \varepsilon$

(ε is the given accuracy; $i = 0, 1, \ldots, n-1$)

$\Rightarrow$ stop and take $y_{i,\,j+1} = y_i(t_j+h)^{(h/2)}$

Otherwise:

Halve the step as the new h (1/2 of the original h), repeat the procedure

↖

....

Do the calculation till $\max_{(i = 0, 1, \ldots, n-1)} | y_{i\,j}^{(h/2)} - y_{i\,j}^{(h)} | < \varepsilon$. The variable step algorithm is of higher accuracy. (in our calculation ε=10^{-6})

2. Adams-Bashforth-Moulton method

The method is of higher accuracy. The algorithm is:

$$\bar{y}_{i,j+1} = y_{ij} + [55f_{ij} - 59f_{i,j-1} + 37f_{i,j-2} - 9f_{i,j-3}]h/24$$

From which, we have:

$$f_{i,j+1} = f_i(t_{j+1}, \bar{y}_{0,j+1}, \bar{y}_{1,j+1}, \cdots, \bar{y}_{n-1,j+1})$$

Insert this formula into the following expression:

$$y_{i,j+1} = y_{ij} + [9f_{i,j+1} + 19f_{ij} - 5f_{i,j-1} + f_{i,j-2}]h/24$$

here

$$f_{ik} = f_i(t_k, y_{0k}, y_{1k}, \cdots, y_{n-1,k}) \quad (k = j\text{-}3, j\text{-}2, j\text{-}1, j)$$

$$(i = 0, 1, \ldots, n\text{-}1)$$

This is a multi-step method. In calculating the functions at the new point, the functions of the previous four points are needed. Runge-Kutta algorithm can be employed for their calculations. The initial four functions are y_{i0}, y_{i1}, y_{i2}, y_{i3} ($i = 0, 1, \ldots, n$-1). y_{i0} is the given initial value.

B. Numerical method for Poincare surface of section

The surface of section (q_+, p_+) is the plot of the trace points as $q_- =$ const (const is a constant which is chosen as 0) and $p_- <0$ or >0. Numerically, $q_- =$ const is hardly realized. What we really care is the change of sign of ($q_- -$ const). As $[q_-(t_{j+1}) - \text{const}] \times [q_-(t_j) - \text{const}] < 0$, ($q_+$, p_+) that is with q_- closest to $q_- =$ const can be obtained by using $q_-(t_{j+1})$ as an independent variable to integrate $q_+(t_{j+1})$, $p_+(t_{j+1})$ back one step. If energy constancy is still in the accuracy limit, then q_+, p_+ are what we want. Explicitly:

Since
$$dq_\pm / dt = \partial H_{eff} / \partial p_\pm$$
$$dp_\pm / dt = -\partial H_{eff} / \partial q_\pm$$

then
$$\frac{dq_+}{dq_-} = \frac{\partial H_{eff} / \partial p_+}{\partial H_{eff} / \partial p_-}$$
$$\frac{dp_+}{dq_-} = -\frac{\partial H_{eff} / \partial q_+}{\partial H_{eff} / \partial p_-}$$

and
$$\frac{q_+^* - q_+(t_{j+1})}{q_-^* - q_-(t_{j+1})} = \frac{\partial H_{eff} / \partial p_+}{\partial H_{eff} / \partial p_-}$$
$$\frac{p_+^* - p_+(t_{j+1})}{q_-^* - q_-(t_{j+1})} = -\frac{\partial H_{eff} / \partial q_+}{\partial H_{eff} / \partial p_-}$$

Take q^*_- = const and substitute in the above expressions. The obtained q^*_+, p^*_+ are the values on the surface of section at q_- = const:

$$q_+^* = \frac{\partial H_{eff} / \partial p_+}{\partial H_{eff} / \partial p_-}[const - q_-(t_{j+1})] + q_+(t_{j+1})$$
$$p_+^* = -\frac{\partial H_{eff} / \partial q_+}{\partial H_{eff} / \partial p_-}[const - q_-(t_{j+1})] + p_+(t_{j+1})$$

They are much more accurate than $q_+(t_j)$ and $p_+(t_j)$, or $q_+(t_{j+1})$ and $p_+(t_{j+1})$!

For details, see : M. Henon, On the numerical computation of Poincare maps, Physica 5D (1982) 412

Chapter 14

Chaotic and periodic motions of DCN

14.1 Chaotic motion of DCN

DCN possesses D–C and C–N bonds. From its spectra (Ref.14.1), it is concluded that there is a series of levels that involve only the bond stretches and are independent of the bend motion. In other words, its stretch motion can be separately treated from the bend.

Consider the system as two coupled Morse oscillators:

$$H_0 = \omega_s(n_s + \frac{1}{2}) + \omega_t(n_t + \frac{1}{2}) + X_{ss}(n_s + \frac{1}{2})^2 + X_{tt}(n_t + \frac{1}{2})^2$$
$$+ X_{st}(n_s + \frac{1}{2})(n_t + \frac{1}{2})$$

Here, ω stands for the harmonic frequency, X the anharmonic coefficient. s and t are for the D–C and C–N stretches, respectively.

The coupling terms can be determined by the fit of the eigenenergies of the algebraic Hamiltonian to the experimental data. The results are that there are two main resonances, 1:1 and 2:3, i.e.,

$$H_{st} = K_{st}(a_s^+ a_t + h.c.)$$

$$H_K = K(a_s^+ a_s^+ a_t a_t a_t + h.c.)$$

The coefficients are listed in Table 14.1

Table14.1 The coefficients (in cm^{-1}) of the algebraic Hamiltonian of DCN

ω_s	ω_t	X_{ss}	X_{tt}	X_{st}	K_{st}	K
2681.4	1948.9	-21.2	-0.3	-34.3	-88.7	13.8

For H_{st}, $P_1 = n_s + n_t$ is conserved. For H_K, $P_2 = n_s/2 + n_t/3$ is conserved. For $H = H_0 + H_{st} + H_K$, no quantities are conserved except energy. Then, the system can be considered as two unbounded (unless close to dissociation) bosons. The algebra is $h_4 = \{a^+, a, a^+a, I\}$ (I is the identity). h_4 has two u(1) subalgebras: $\{a^+a\}$ and $\{I\}$. Hence, the coset space of DCN vibrational system is the direct product of two $H(4)/\mathrm{U}(1)\otimes\mathrm{U}(1)$. Since the system is unbounded under this algebraic approximation, for the purposes of our calculation, we take $n_s + n_t \le 10$ for which there are 66 levels. Since they are far below dissociation, the approximation is reasonable.

By Heisenberg's correspondence (Section 2.6), in the coset space (q_s, p_s, q_t, p_t), we have:

$$n_s = (q_s^2 + p_s^2)/2, \quad n_t = (q_t^2 + p_t^2)/2$$

$$H_{st} = K_{st}(q_s q_t + p_s p_t)$$

$$H_K = K[(q_s^2 - p_s^2)(\sqrt{2}q_t^3 - \frac{3n_t}{\sqrt{2}}q_t) - 2q_s p_s(\sqrt{2}p_t^3 - \frac{3n_t}{\sqrt{2}}p_t)]$$

Hamilton's equations are:

$$\partial H/\partial p_\alpha = dq_\alpha/dt, \quad \partial H/\partial q_\alpha = -dp_\alpha/dt \qquad (\alpha = \mathrm{s},\ \mathrm{t})$$

For these 66 levels, we can obtain their solution space from $H(q_\alpha, p_\alpha) = E$ with E the eigenenergy. Then, we randomly choose 200 initial points to calculate the trajectories. Since the system energy is conserved and positive and negative Lyapunov exponents are paired, we need only to calculate the maximal one, λ. See Chapter13 for its calculational details.

For a trajectory, if $\lambda > 0$, then we call it chaotic. If $\lambda = 0$, then it is regular. For each level's energy, we calculate 200 λ's and take their average: $\langle\lambda\rangle$. The magnitude of $\langle\lambda\rangle$ will show the degree of chaoticity of an eigenstate. The result is shown in Fig. 14.1. The levels can be divided into three categories: Levels 1 to 15 (2031cm^{-1} to 11915cm^{-1}) are regular. Levels 16 to 35 (12502cm^{-1} to 18246cm^{-1}) show significant increase of chaoticity. In this energy range, chaoticity and regularity coexist. For levels above 36, all the trajectories are chaotic. The chaoticity increases rather smoothly. We will address the physics behind this observation in Chapter 19.

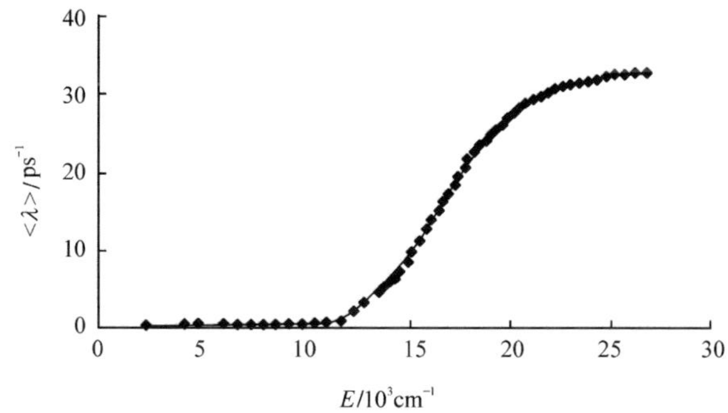

Fig.14.1 The averaged Lyapunov exponents $\langle\lambda\rangle$ for the eigenlevels of DCN

14.2 Periodic trajectories

Generally speaking, nonlinearity induces chaos. However, it will also induce periodicity. The induced periodicity is different from the periodicity by linearity with peculiar properties. In a nonlinear system, chaotic and periodic trajectories are closely related to each other. According to the KAM theorem (Chapter 3), nonintegrability of highly excited vibration will destroy the periodic and quasi-periodic trajectories due to inter-mode couplings with the formation of new (quasi-)periodic and chaotic trajectories in the phase space. Gutzwiller has pointed out that semiclassically, quantal density of states is related to the remnant periodic trajectories. Hence, periodic trajectories are related to quantization for a nonintegrable system.

DCN is such a case. At high energy, its vibrational phase space is full of chaotic *sea* among which there are numerous periodic trajectories. These periodic trajectories are different from the classical harmonic motion. We will introduce their characteristics later. Noteworthy is the appearance of period-3 trajectories. From Sarkovskii, we know that the appearance of a period-3 trajectory implies the existence of chaos. Hence, the system is chaotic. (See Section 3.3).

What needs to be emphasised is that these periodic trajectories are the induction of nonlinearity. They are due to the nonlinear effect and are essentially different from the linear periodic trajectories. For a chaotic system, periodic trajectories are crucially important. They, together with chaotic trajectories, form the basics of the system.

In the following, we will address the period-1, 5, 3, 7, 8, 9, 12, 15 and 18 trajectories. For convenience, the periods are defined by the observation on the Poincare surface of a section (q_s, p_s) $(q_t = 0, p_t < 0)$. As observed from another surface of sections such as (q_t, p_t) $(q_s = 0, p_s < 0)$, the periods may vary. But this is not essential physically.

14.2.1 Period-1, 5 trajectories

In the 11th levels, there are two period-1 trajectories. Each trace on the (q_s, p_s) and (q_t, p_t) surfaces spans a deviation less than 5×10^{-3} in 10 ps. We will stick to this tolerance in the following discussion. Fig.14.2 (a) and (b) show their phases. ($\phi_\alpha = \tan^{-1}(-p_\alpha / q_\alpha)$, $\alpha = s,t$), respectively. In Fig. 14.2(a), ϕ_s and ϕ_t are in phase and the mode is symmetric. Fig. 14.2(b) shows the antisymmetric mode since ϕ_s and ϕ_t are out of phase. Their periods are 17.2 fs and 10 fs, respectively.

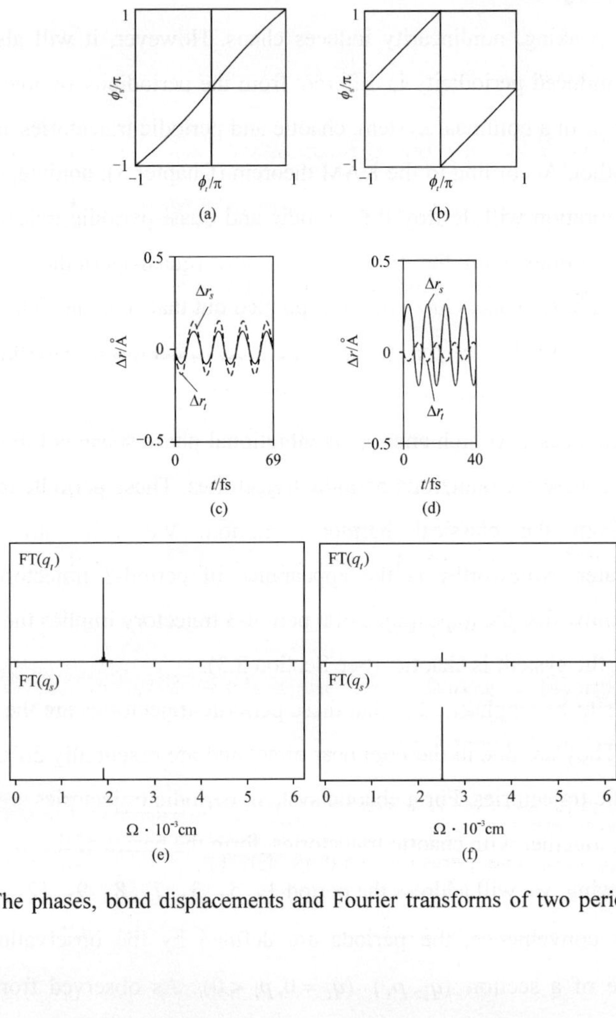

Fig.14.2 The phases, bond displacements and Fourier transforms of two period-1 trajectories in level 11

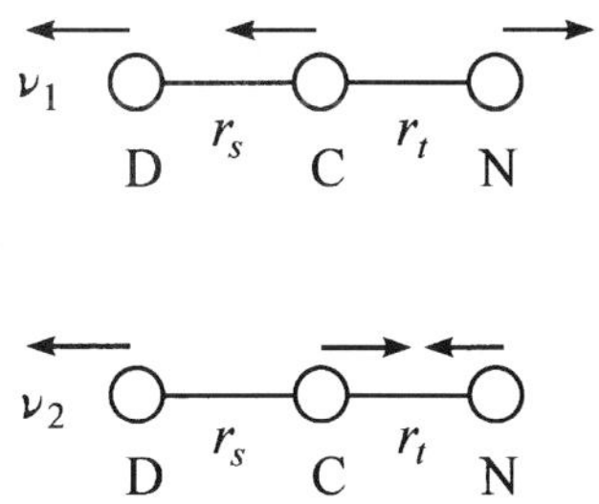

Fig.14.3 ν_1 and ν_2 normal modes

The bond displacement Δr corresponding to (q, p) is obtained through the following formula (Section 1.2) if we consider both D–C and C–N as Morse oscillators:

$$\Delta r = a^{-1} \ln\{[1-(1-\lambda^2)^{1/2} \cos\phi]/\lambda^2\}$$

Here, $a = (-2X\mu)^{1/2}$, μ the reduced mass, $\lambda = 1-(2n+1)/k$, $k = \omega/(-X)$, $n = (q^2+p^2)/2$. Fig.14. 2 (c) and (d) are the evolving Δr_s and Δr_t. The mode character, symmetry or antisymmetry, is vivid. Fig.14.3 shows ν_1 and ν_2 normal modes. ν_1 is symmetric while ν_2 is antisymmetric. The amplitude ratios A_s/A_t for these period-1 trajectories are 0.63 and 6.37 which are quite different from those of ν_1 and ν_2: 0.21 and 1.44. The action (proportional to the square of amplitude) ratios of D–C and C–N are 0.04:1 and 2.07:1 for ν_1 and ν_2. Fig.14.2(e) and (f) show the Fourier transforms of their q_s and q_t. For the symmetric and antisymmetric period-1 trajectories, there are peaks at 1914 cm^{-1} and 2528 cm^{-1}, respectively. These are different from ν_1, ν_2, which are at 1925 cm^{-1} and 2630 cm^{-1}.

In the 23rd level, we observe a symmetric period-5 trajectory. Fig.14.4(a) is the relation of its ϕ_s, ϕ_t. The period is 85.0 fs. On (q_s, p_s) and (q_t, p_t) surfaces, there are five points. The Fourier transforms of q_s and q_t also show only one peak at 1914 cm^{-1}. This peak is broader than those of period-1 trajectories. This is not unexpected since their periods are different. We also expect an antisymmetric period-5 trajectory in analogy with the period-1 (and period-3, 5, 7 to be introduced later) trajectory at 2528 cm^{-1}, though we did not find it.

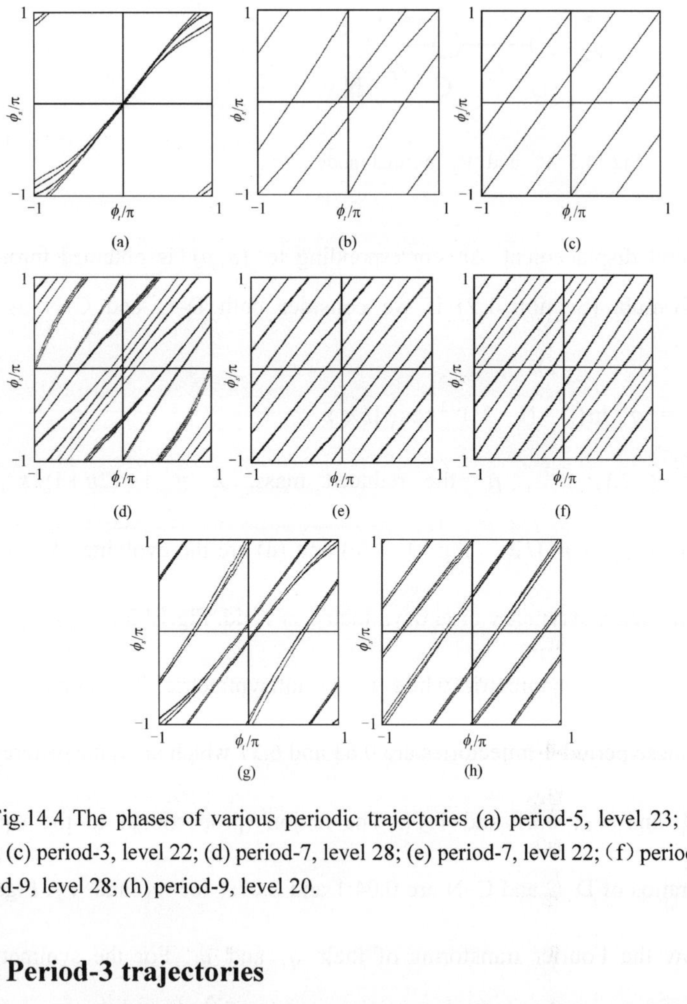

Fig.14.4 The phases of various periodic trajectories (a) period-5, level 23; (b) period-3, level 22; (c) period-3, level 22; (d) period-7, level 28; (e) period-7, level 22; (f) period-8, level 15; (g) period-9, level 28; (h) period-9, level 20.

14.2.2 Period-3 trajectories

Fig.14.5(a) shows that on the (q_s, p_s) surface there are two period-3 trajectories in level 22. One is in the *inner* region and the other is in the *outer* region. (A similar situation appears in period-1, 7, 9 trajectories.) Fig.14. 5 (b) shows what is observed on the (q_t, p_t) surface. (Since there are four points on the surface, we may consider it of period 4.) Fig.14. 4(b) and (c) show their phase relations. In a period, Δr_s has 4 periods, Δr_t has 3 periods. Their periods are 56 fs and 52 fs, respectively.

In a period, there is time interval, t_s, in which Δr_s and Δr_t are in phase and the interval, t_a, in which Δr_s and Δr_t are out of phase. For the inner period-3 trajectory of the 11th level, t_s/t_a is 1.69. This value increases to 1.96 for the 30th level. This shows that period-3 trajectories are composed of v_1 and v_2 with a ratio of 2:1. (Note that Δr_s, Δr_t of v_1 and v_2 are in and out of phase, respectively.) The action ratio on D–C and C–N is 2.15:3 ($2\times 0.04+1\times 2.07:2\times 1+1\times 1$, see Section14.2.1). For higher levels, the component of v_1 in period-3 trajectory increases. For the outer period-3 trajectory, t_s/t_a drops from 1.08 of level 11 to 0.96 of level 22. Hence, this period-3 trajectory is composed of v_1 and v_2 with a ratio of 1:1. Indeed, the action ratio, n_s/n_t, on D–C and C–N is close to 1 ($1\times 0.04+1\times 2.07$: $1\times 1+1\times 1$, see Section 14.2.1). The Fourier spectrum also shows that inner period-3 trajectory has two peaks at 1914 cm^{-1} and 2553 cm^{-1} while for the outer period-3 trajectory, the peaks are at 1878 cm^{-1} (Ω_t) and 2503 cm^{-1} (Ω_s) and there are some other smaller peaks: 3191 cm^{-1} ($2\Omega_s-\Omega_t$) and 1253 cm^{-1} ($2\Omega_t-\Omega_s$). These peaks are different from v_1 and v_2. From Sarkovskii, we know that the appearance of period-3 trajectories implies that D–C motion is chaotic.

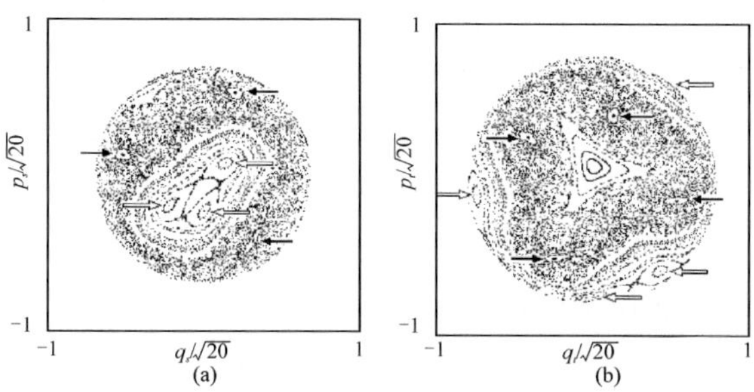

Fig.14.5 Period-3 trajectories of level 22 on the (q_s, p_s) and (q_t, p_t) surfaces. Arrows show their locations.

14.2.3 Period-7, 8 trajectories

The period-7 trajectory as shown on the (q_s, p_s) surface is of period 9 if viewed on the (q_t, p_t) surface. The phase relations of the inner period-7 trajectory of level 28 and the outer period-7 trajectory of level 22 are shown in Fig.14.4(d),(e). Though their phase relations are different, their periods are close to 119 fs and t_s/t_a values are 1.53 and 1.08, respectively. The Fourier spectra of q_s and q_t show peaks at 1878 cm^{-1}, 2418 cm^{-1} and 1927 cm^{-1}, 2479 cm^{-1} together with more complicated minor components.

The period-8 trajectory on (q_s, p_s) becomes period-11 on (q_t, p_t). For instance, the period-8 trajectory of level 15 possesses a period of 148 fs. The phase relation is shown in Fig.14.4(f). t_s/t_a is 0.83. Frequency peaks are at 1792 cm^{-1} and 2467 cm^{-1}.

These period-9, 11 trajectories on the (q_t, p_t) surface, according to Sarkovskii theorem, also imply chaos of the C–N stretch.

14.2.4 Period-9, 12, 15, 18 trajectories

The trajectories of periods that are multiples of 3 (like 9, 12, 15 and 18) possess characteristics similar to a period-3 trajectory. On the (q_s, p_s) surface, the corresponding 9, 12, 15 and 18 points aggregate into three clusters of 3, 4, 5 and 6 points, respectively. On (q_t, p_t), there are 4 clusters, and each possesses 3, 4, 5 and 6 points, respectively. Hence, from the (q_t, p_t) surface, the periods are 12, 16, 20 and 24. Shown in 14.4(g) and (h) are the phase relations of the inner period-9 trajectory of level 28 and the outer period-9 trajectory of level 20. Indeed, they are *similar* to those of period-3 trajectories. This is the same for period-12, 15 and 18 trajectories. Their periods are longer: 154 fs and 167 fs for period-9 trajectories, 206 fs for period-12 trajectory, 281 fs for period-15 trajectory and 326 fs for the period-18 trajectory. Their t_s/t_a values are:1.92, 0.95; 1.89; 0.95; 0.98, respectively. Frequency spectra of q_s and q_t show peaks at: 1939 cm^{-1}, 2589 cm^{-1}; 1792 cm^{-1}, 2393 cm^{-1}; 1927 cm^{-1}, 2577 cm^{-1}; 1779 cm^{-1}, 2368 cm^{-1} and 1841 cm^{-1}, 2454 cm^{-1} together with minor complicated combinations. As a period increases, the trajectory also approaches chaos and the frequency spectrum becomes more complicated as shown in Fig.14.6.

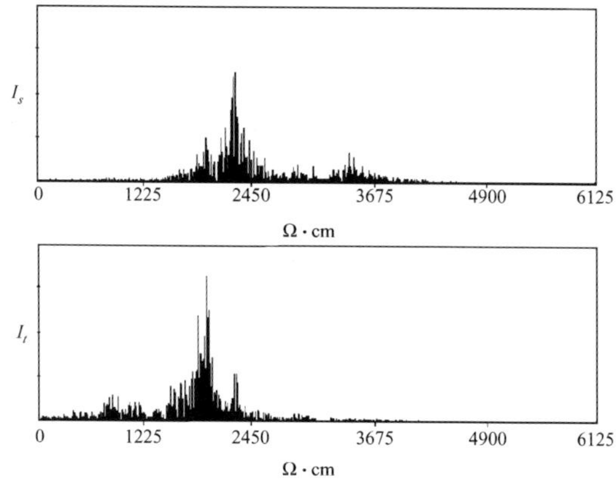

Fig.14.6 Frequency spectra of a chaotic trajectory

Just like v_1, in-phase of Δr_s and Δr_t implies that more the C–N stretch is involved and the larger t_s/t_a is. Conversely, like v_2, out-of-phase of Δr_s and Δr_t implies that more the D–C stretch is involved and the smaller t_s/t_a is. For the period-3 trajectory of level 15, t_s/t_a=1. Therefore, if a trajectory is in the outer region of the period-3 trajectory, then its t_s/t_a<1. Otherwise, it is in the inner region and its t_s/t_a>1. Chaotic trajectories are in the outer region of the period-3 trajectory with t_s/t_a in the range of 0.7-0.95. In other words, chaotic trajectories involve more D–C motion.

However, close to the center of the (q_t, p_t) surface as shown in Fig.14.5(b), there is evidence that energy of high excitation is localized on the quasi-periodic mode of the D–C stretch. This mode is on the outer rim of the (q_s, p_s) surface. Fig. 14.7 shows the the relative distribution of actions on D–C (n_s) and C–N (n_t) of this mode. Indeed, action is concentrated on D–C and highly localized.

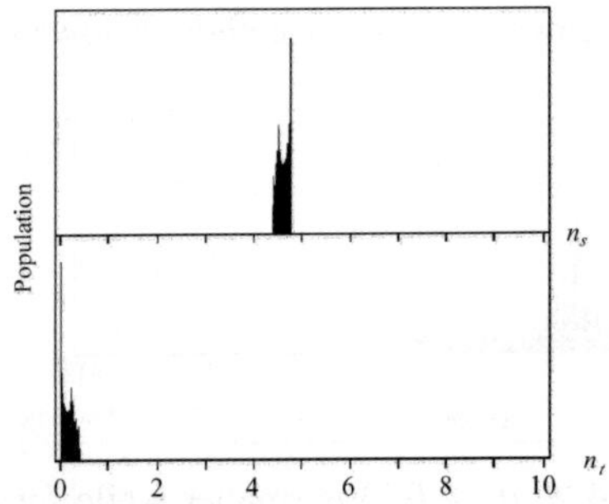

Fig.14.7 The relaive distribution of actions on D–C (n_s) and C–N (n_t), showing that action is concentrated on D–C and highly localized. See text for details.

In summary, regardless of the period, periodic trajectories are composed of ν_1 and ν_2 in a ratio of 2:1 or 1:1 and the action ratios on D–C and C–N are 2:3 and 1:1. These are just in conformity with the resonance forms. In other words, the contents of periodic trajectories are decided by the resonance forms. Table 14.2 summarizes this observation.

Table14.2 Periodic trajectories are composed of ν_1 and ν_2 in a ratio of 2:1 or 1:1 and the action ratios on D–C and C–N are 2:3 and 1:1. See text for details.

Periodic trajectory	t_s/t_a	compositions	$n_s:n_t$	$n_s:n_t$ (truncated)
p^3(outer)	1: 1	$\nu_1+\nu_2$	2.11:2	1: 1
p^3(inner)	2: 1	$2\nu_1+\nu_2$	2.15 : 3	2: 3
p^7(outer)	1.08: 1	$1.08\nu_1+\nu_2$	2.11:2.08	1: 1
p^7(inner)	1.53: 1	$1.53\nu_1+\nu_2$	2.13:2.53	~1: 1
p^8	0.83: 1	$0.83\nu_1+\nu_2$	2.10:1.83	~1: 1
p^9(outer)	0.95: 1	$0.95\nu_1+\nu_2$	2.11:1.95	1: 1
p^9(inner)	1.92: 1	$1.92\nu_1+\nu_2$	2.14:2.92	2: 3
p^{12}	1.89: 1	$1.89\nu_1+\nu_2$	2.14:2.89	~2: 3
p^{15}	0.95: 1	$0.95\nu_1+\nu_2$	2.11:1.95	1: 1
p^{18}	0.98: 1	$0.98\nu_1+\nu_2$	2.11:1.98	1: 1

In the next section, more analysis will be offered to the origin of the chaotic motion of the system due to the D–C stretch.

14.3 Chaotic motion originating from the D–C stretch

The phase spaces for levels from 16 to 30 can be divided into two parts: the inner portion corresponding to regular motion and the outer portion to the chaotic motion. The relative area ratio is fixed. Fig.14.8 shows Poincare surfaces of section of some regular (in the inner portion) and chaotic (in the outer portion) trajectories of level 19 and their relative n_s, n_t distributions. For the chaotic portion, n_s ($4.4 > n_s > 1.9$) is larger and n_t ($3.4 > n_t > 0.2$) is smaller while for the regular portion, n_s ($1.2 > n_s > 0$) is smaller and n_t ($5.8 > n_t > 4.3$) is larger. This shows that the excitation of D–C (action>1.9), not that of C–N, is the origin of the chaotic motion. Of course, the amount of energy excited is related to chaos. But the situation is not so simple. Energy or action distribution among the modes can also be an important factor for chaotic motion to appear.

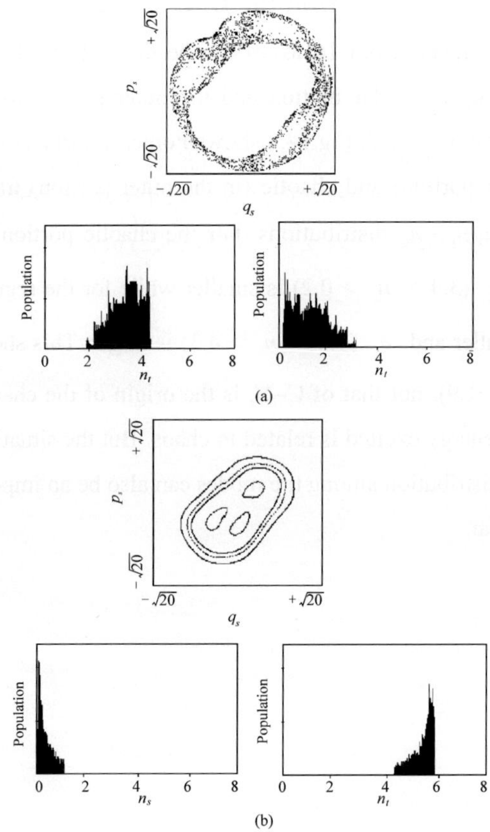

Fig.14.8 Poincare surfaces of section (q_s, p_s) $(p_t < 0, q_t = 0)$ of some chaotic trajectories (a) and regular trajectories (b) of level 19 of DCN and their relative n_s, n_t distributions.

References

14. 1 J. E. Baggott, G. L. Caldow and I. M. Mills, *J. Chem. Soc. Faraday Trans.*, 2 84 (1988)1407.

14. 2 M. C. Gutzwiller, *Chaos in Classical and Quantum Mechanics in Interdisciplinary Applied Mathematics*, Springer, Berlin,1990

14. 3 D. Zheng and G. Wu, *Chem. Phys. Lett.*, 352 (2002) 85.

Chapter 15

Regular classification of highly excited vibrational levels and its physical background

15.1 Introduction: algebraic method

One characteristic of molecular highly excited vibration is its immense number and high density of states. These quanta are bosons and can populate the various modes and transfer among them. Due to the interaction among the quanta, many quantum numbers are destroyed. The destruction of quantum numbers, in another aspect, implies that their classical analogues, the actions which are continuous variables, have to be employed for the description of highly excited vibration. In classical mechanics, angle is a companion of action. Hence, in the study of molecular highly excited vibration, both action and angle are important physical quantities.

Albeit at high excitation, most of the quantum numbers are destroyed and are no longer constants of motion, there are still some good quantum numbers left. These remaining quantum numbers are called the *polyad* numbers. We note that polyad numbers are only *approximate*. They are, in fact, conserved only in certain dynamical processes. This is because as time elapses, the dynamics will become very complicated and the resonance forms may vary so that the polyad number as an operator may no longer commute with the full Hamiltonian.

For instance, formaldehyde has six vibrational modes among which the out-of-plane vibration is quite decoupled from the remaining five in-plane modes. We therefore concentrate on the vibrational system composed of these five in-plane modes. These modes are two equivalent C–H stretches, denoted by the subscripts 1, 5, the C–O stretch by the subscript 2, scissor bending of CH_2 by the subscript 3 and the C–O wagging by the subscript 6. The couplings among these five modes can be represented by the algebraic Hamiltonian:

$$K_{15}(a_1^+a_5 + h.c.) + K_{23}(a_2^+a_3 + h.c.) + K_{36}(a_3^+a_6 + h.c.) + K_{26}(a_2^+a_6 + h.c.)$$

$$+K_{122}(a_1^+a_2a_2+a_5^+a_2a_2+h.c.)+K_{133}(a_1^+a_3a_3+a_5^+a_3a_3+h.c.)$$

$$+K_{166}(a_1^+a_6a_6+a_5^+a_6a_6+h.c.)$$

Here, $h.c.$ stands for Hermitian conjugate. K_{ij} are the coupling strengths.

For these resonances, actions n_1, n_2, n_3, n_5, n_6 are destroyed and are no longer good quantum numbers. However, it is easy to deduce that

$$P=2n_1+n_2+n_3+2n_5+n_6$$

is a conserved quantity *intuitively* by the orders of the creation and destruction operators in the coupling Hamiltonian. It can also be checked by direct calculation that the coupling Hamiltonian commutes with P as operators. For a given P, there are many associated states. Like for $P = 8$, there are 176 states. All these 176 states share a common quantum number $P = 8$.

These five modes can be considered as coupled Morse oscillators with resonances. The part of the Hamiltonian for the Morse oscillators can be written as:

$$H_0=\omega_1(n_1+n_5+1)+\omega_2(n_2+1/2)+\omega_3(n_3+1/2)+\omega_6(n_6+1/2)$$
$$+X_{11}[(n_1+1/2)^2+(n_5+1/2)^2]+X_{22}(n_2+1/2)^2+X_{33}(n_3+1/2)^2+X_{66}(n_6+1/2)^2$$
$$+X_{12}(n_1+n_5+1)(n_2+1/2)+X_{13}(n_1+n_5+1)(n_3+1/2)$$
$$+X_{15}(n_1+1/2)(n_5+1/2)+X_{16}(n_1+n_5+1)(n_6+1/2)$$
$$+X_{23}(n_2+1/2)(n_3+1/2)+X_{26}(n_2+1/2)(n_6+1/2)+X_{36}(n_3+1/2)(n_6+1/2)$$

The advantage of this formulation is its clear physical implication. Each term denotes a concrete physical mechanism. Then, the question that remains is: how can the coefficients ω, X, K be determined?

For instance, for the case of $P = 8$, with various experiments it is plausible to determine all the 176 states (It may not matter much even if not all the 176 states are found.). Then, we try to elucidate ω, X, K by fitting the eigenenergies of this algebraic Hamiltonian to the experimental values via nonlinear routines such as the Marquardt method. In this process, including the assignment of the 176 states, the elucidation of the resonance forms cannot be completed in an *once for all* way. Instead, many trial-and-error procedures are expected. In the literature, for formaldehyde, there were such reports as tabulated in Table 15.1 (see Refs. 15.1, 15.2).

Table 15.1 The coefficients (cm^{-1}) of algebraic Hamiltonian for formaldehyde. See text for details.

ω_1	2885.66
ω_2	1735.45
ω_3	1539.46
ω_6	1264.90
K_{15}	-36.05
K_{23}	-62.22
K_{26}	-0.14
K_{36}	86.47
K_{122}	9.23
K_{133}	2.28
K_{166}	3.61
X_{11}	-56.94
X_{12}	0.03
X_{13}	-22.32
X_{15}	-0.51
X_{16}	119.36
X_{22}	-11.89
X_{23}	-1.66
X_{26}	37.50
X_{33}	-19.82
X_{36}	27.76
X_{66}	-18.86

How do we obtain the eigenenergies of this algebraic Hamiltonian? For instance, for $P = 8$, we can choose 176 bases of $|n_1, n_2, n_3, n_5, n_6>$ by which a Hamiltonian matrix of 176×176 can then be constructed. By diagonalizing it, the 176 eigenenergies and states can be elucidated. Each eigenstate $|\varphi>$ is a linear combination of the basis states :

$$|\varphi> = \sum C|n_1, n_2, n_3, n_5, n_6>$$

$|n_1, n_2, n_3, n_5, n_6>$ are called the zeroth-order eigenstates. This is because they are the eigenstates as all K_{ij} are zero (i.e., without any resonances). In this situation, n_1, n_2, n_3, n_5, n_6 are good quantum numbers.

15.2 Diabatic correlation, formal quantum numbers and ordering of levels

Shown in the column to the right in Fig.15.1(a) are the 176 levels of formaldehyde with $P = 8$. No ordering of the levels is evident. In the following, we will try to seek the ordering of the levels. For this, the retrieval of the approximate quantum numbers or constants of motion is the crucial step. We first introduce the concept of diabatic correlation.

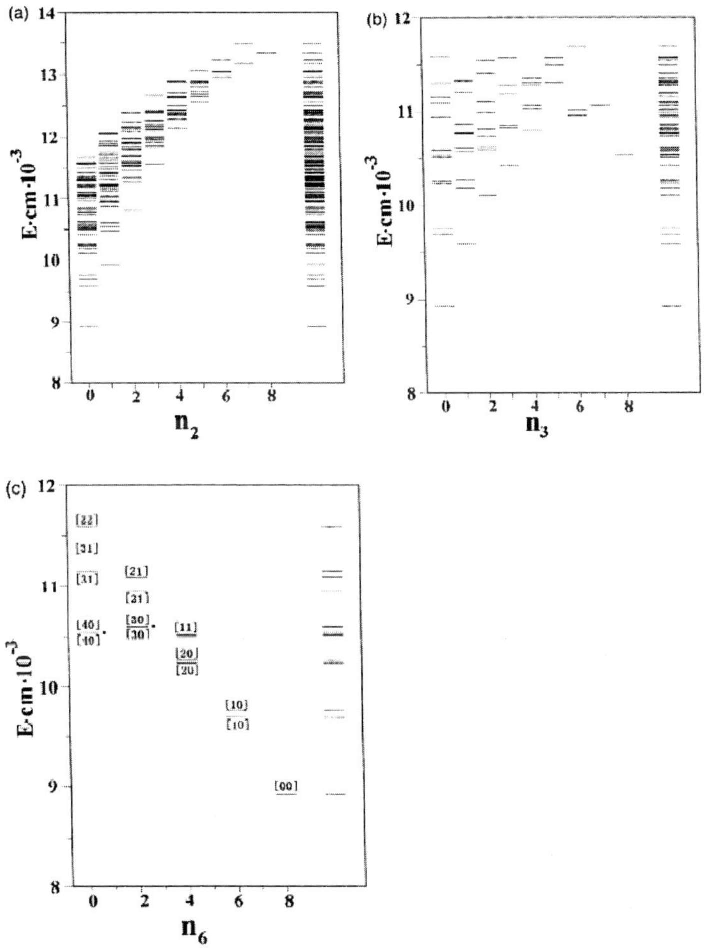

Fig.15.1 The 176 levels of formaldehyde with $P=8$ and their ordering by various actions. · denotes that the energy spacing is very small.

In quantum mechanics, it is well known that as the interaction between two levels varies, the levels will avoid crossing each other. Along the noncrossing route is the so-called adiabatic correlation, while along the route which leads to crossing is the diabatic correlation.

For the seven resonances of formaldehyde, we can switch them off one by one. Correspondingly, there is avoidance of crossing. At each of these points we stick to the diabatic correlation. This will finally lead to the correlation of each level of the full Hamiltonian with a zeroth-order level of H_0. The eigenfunctions of H_0 are $|n_1, n_2, n_3, n_5, n_6>$. Therefore, formally, we can label the levels (of the full Hamiltonian) with the correlated quantum numbers of $|n_1, n_2, n_3, n_5, n_6>$. This labeling is just formal since due to resonances, they are no longer good quantum numbers or constants of motion for the levels of the full Hamiltonian. For convenience, we call $[n_1, n_2, n_3, n_5, n_6]$ the *formal quantum numbers*. (For the eigenstates of H_0, they are indeed the good quantum numbers.) Fig.15.2 shows part of the switching-off of the seven resonances. (Note that $2n_1 + n_2 + n_3 + 2n_5 + n_6 = 8$) Also shown are the labelings of the levels by $[n_1, n_2, n_3, n_5, n_6]$ and the diabatic correlation.

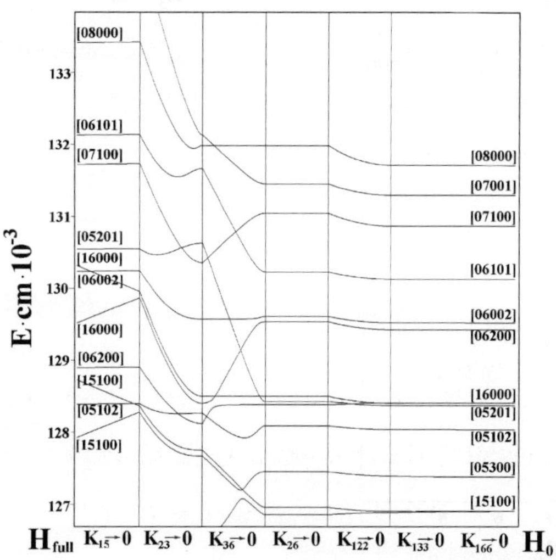

Fig.15.2 The switching-off of the seven resonances of formaldehyde and the labelings of the levels by $[n_1, n_2, n_3, n_5, n_6]$ via the diabatic correlation. H_0 is the Hamiltonian without resonances. H_{full} is the full Hamiltonian.

Now we can reconstruct the level diagram by the formal quantum numbers. Fig.15.1(a) is the reconstructed levels by $n_2 = 0,1,\ldots,8$. No apparent ordering is observed therein. The column to the right in Fig.15.1(b) shows the levels with $n_2 = 0$ and the rest are the reconstructions of these levels by $n_3 = 0,1,\ldots,8$. Similarly, the column to the right in Fig.15.1(c) shows the levels with $n_3 = 0$ ($n_2 = 0$) and the rest are the reconstructions by $n_6 = 0, 2, 4, 6, 8$ in which each level is labeled by $[n_1\ n_5]$, corresponding to the formal quantum numbers of the two C–H stretches. Then, the regular pattern is obvious. For each n_6 column, lower levels are almost degenerate, showing the local mode character. While for the higher levels, the nearest level spacings are almost the same, showing the character of the normal mode. These are the typical characteristics of a two-mode system (For details, see Chapter 4). In Fig.15.1, only the ordering of the levels by $n_2 = 0$, $n_3 = 0$ is shown. The others can also be reconstructed to show regular patterns.

The above analysis shows that formal quantum numbers are indeed useful though they are not good quantum numbers or constants of motion in the strict sense. By the formal quantum numbers, levels which appear *irregularly* can be reconstructed to show regular patterns. This is very useful for the assignment and classification of highly excited vibrational levels. Due to couplings, most of the quantum numbers or the constants of motion of the highly excited vibration are destroyed. This causes the difficulty in the assignment and classification of its levels. The reconstruction by the formal quantum numbers shows a new route for this topic.

15.3 Acetylene case

Acetylene possesses five in-plane motions. They are symmetric and antisymmetric stretches of C–H (subscripts 1, 3), C–C stretch (subscript 2) and *trans*, *cis* bends of C–H (subscripts 4,5). These modes have the following anharmonicities and resonances (Ref.15.3):

$$\sum\omega_i n_i + \sum X_{ij} n_i n_j + 1/4K_{4455}(a_4^+ a_4^+ a_5 a_5 + h.c.)$$
$$-1/8K_{3245}(a_3^+ a_2 a_4 a_5 + h.c.) - 1/4K_{1244}(a_1^+ a_2 a_4 a_4 + h.c.)$$
$$-1/4K_{1255}(a_1^+ a_2 a_5 a_5 + h.c.) - 1/8K_{1435}(a_1^+ a_4^+ a_3 a_5 + h.c.)$$
$$+1/4K_{1133}(a_1^+ a_1^+ a_3 a_3 + h.c.)$$

Table15.2 Coefficients (cm^{-1}) of the algebraic Hamiltonian for acetylene. See text for details.

ω_1	3398.74
ω_2	1981.71
ω_3	3316.09
ω_4	609.02
ω_5	729.17
K_{4455}	-11.0
K_{3245}	-18.28
K_{1244}	6.38
K_{1255}	6.38
K_{1435}	29.04
K_{1133}	105.83
X_{11}	-26.57
X_{12}	-12.62
X_{13}	-105.09
X_{14}	-15.58
X_{15}	-10.85
X_{22}	-7.39
X_{23}	-6.10
X_{24}	-12.48
X_{25}	-1.57
X_{33}	-27.74.
X_{34}	-6.96
X_{35}	-8.69
X_{44}	3.08
X_{45}	-2.41
X_{55}	-2.34

The coefficients are tabulated in Table 15.2. For this Hamiltonian, the polyad number is $P = 5n_1 + 3n_2 + 5n_3 + n_4 + n_5$. This is easily checked by the orders of the creation and destruction operators in the coupling Hamiltonian. Fig.15.3 shows the switching-off of the resonances and the diabatic correlation from the full Hamiltonian to the zeroth-order Hamiltonian. By this procedure, we can label the levels of the full Hamiltonian with a set of formal quantum numbers. The column to the right in Fig.15.4 shows the 134 levels with P = 15. In the second column are the 51 levels with the formal quantum numbers $n_1 = n_3 = 0$. No apparent ordering is observed. The next step is the reconstruction of these 51 levels with n_2 = 0,1,..,5. n_2 = 0 corresponds to a pure C–H bend and are denoted by $[n_4, n_5]$. [15, 0] represents the *trans* bend which is lower than the *cis* bend [0, 15]. Lower levels are almost degenerate while higher levels show almost equal spacings. For nonzero n_2, the C–C stretch will affect the bends (like via the resonances K_{3245}, K_{1244}, K_{1255}). The spacings are nearly equi-distant.

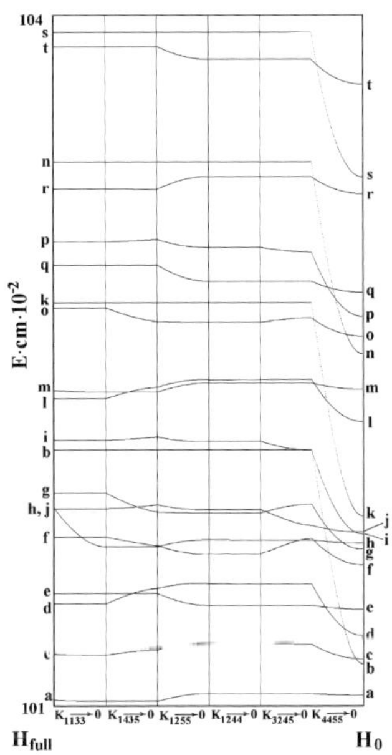

Fig.15.3 The switching-off of the resonances of Hamiltonian H_{full} and the diabatic correlation to the zeroth-order Hamiltonian H_0 for acetylene

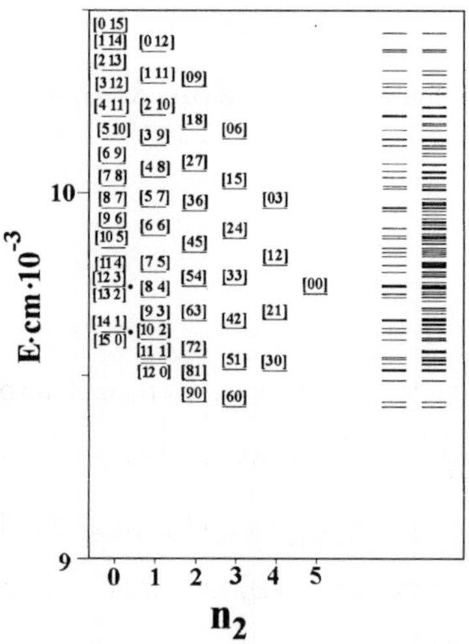

Fig.15.4 The reconstruction of the levels of acetylene with $n_2 = 0,1,..,5$ by $[n_4, n_5]$. · denotes that the energy spacing is very small. The column on the far right shows the 134 levels with $P = 15$. In the second column are the 51 levels with the formal quantum numbers $n_1 = n_3 = 0$. See text for details.

The cases of formaldehyde and acetylene show that for the highly excited vibration, by the diabatic correlation, we can employ the quantum numbers of the zeroth-order Hamiltonian, called the formal quantum numbers, to reconstruct its levels in a very regular pattern. Although, due to resonances, these formal quantum numbers are no longer the constants of motion, they are still useful for the ordering of the levels. This means that formal quantum numbers are not completely meaningless quantities.

15.4 Background of diabatic correlation

In the following, we will show that besides polyad numbers, there are other approximately conserved quantities. First, we explain the physical background of diabatic correlation. The tri-atomic systems of H_2O, D_2O, H_2S are taken as examples. Their *algebraic* Hamiltonian is : (See Chapter 7):

$$H = H_0 + H_{st} + H_F$$

$$H_0 = \omega_s(n_s + n_t + 1) + \omega_b(n_b + \frac{1}{2}) + X_{ss}[(n_s + \frac{1}{2})^2 + (n_t + \frac{1}{2})^2]$$
$$+ X_{bb}(n_b + \frac{1}{2})^2 + X_{st}(n_s + \frac{1}{2})(n_t + \frac{1}{2}) + X_{sb}(n_s + n_t + 1)(n_b + \frac{1}{2})$$

$$H_{st} = K_{st}(a_s^+ a_t + h.c.)$$

$$H_F = K_{sb}(a_s^+ a_b a_b + a_t^+ a_b a_b + h.c.)$$

where H_0 includes the nonlinear effects, H_{st} is the 1:1 and 2:2 resonances of *s*, *t* stretches, H_F is the Fermi resonance among the bend *b* and *s*, *t* stretches, ω, X are the mode frequency and anharmonicity, K is the coupling strength. By fitting the eigenenergies of H to the experimental observations, these coefficients can be determined. They are listed in Table 7.1. The polyad number is $P = n_s + n_t + n_b/2$.

Eigenstate ψ can be expressed as a linear combination of the eigenstates $|i_s j_t k_b\rangle$ of H_0 (That notations i_s, j_t, k_b are used instead of n_s, n_t, and n_b is just to stress that they are the quantum numbers for H_0. Without confusion, we can still employ the notation n_s, n_t, n_b). Since $i_s + j_t + k_b/2$ is a constant, index k_b can be omitted. So we have:

$$\psi = \sum C_{i_s j_t} |i_s j_t\rangle$$

Fig.15.5(a) and (b) show $|C_{i_s j_t}|^2$ of the 40th (L40) and 42 nd (L42) levels of H_2O with $P = 10$. The Hamiltonian is $H_0 + H_{st} + \lambda H_F$, with $\lambda = 0.36$, 0.44. The coordinate is the notation of $|i_s j_t\rangle$ and is expressed as $\langle i_s, j_t\rangle$. Due to interaction, these two levels repulse each other as shown in Fig.15.5(c). From Fig.15.5(a) and (b), the distributions of $|C_{i_s j_t}|^2$ of these two levels change considerably as λ varies from 0.36 to 0.44. However, if the diabatic correlation is followed, the patterns of the distributions are preserved. Hence, for the eigenstates of *H*, during diabatic correlation, the pattern of the distributions of $|C_{i_s j_t}|^2$ is more or less preserved. This means that formal quantum numbers are related to the conservation in the system.

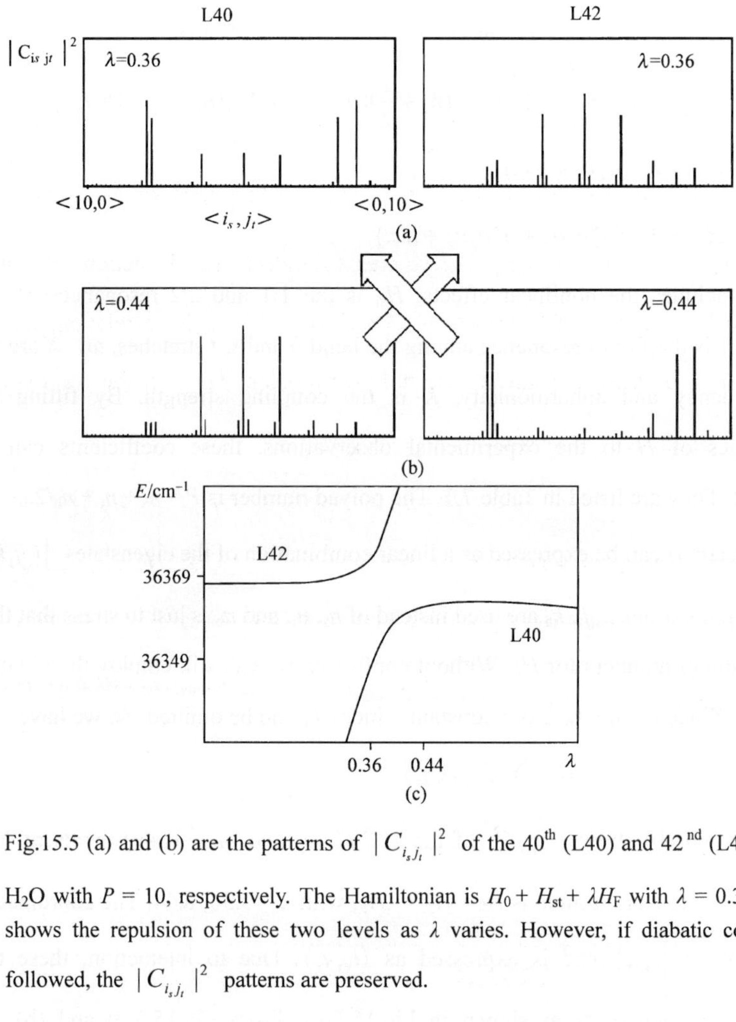

Fig.15.5 (a) and (b) are the patterns of $|C_{i_s j_t}|^2$ of the 40th (L40) and 42nd (L42) levels of H_2O with $P = 10$, respectively. The Hamiltonian is $H_0 + H_{st} + \lambda H_F$ with $\lambda = 0.36, 0.44$. (c) shows the repulsion of these two levels as λ varies. However, if diabatic correlation is followed, the $|C_{i_s j_t}|^2$ patterns are preserved.

15.5 Approximately conserved quantum numbers

We can obtain the approximate quantum numbers by switching off H_{st} and H_F alternatively.

(1) Consider $H_0+H_{st}+\lambda_2 H_F$, with λ_2 varying from 1 to 0. As $\lambda_2 = 0$, the eigenstates of H_0+H_{st} possess the exact quantum number n_b. As soon as the eigenstates of H_0+H_{st} are obtained, their n_b's are known. By diabatic correlation, we can correlate the eigenstates of $H_0+H_{st}+ H_F$ with those of H_0+H_{st}. Since the effect of H_F is small, n_b is the approximate quantum number of the full Hamiltonian ($H_0+H_{st}+H_F$).

(2) Consider $H_0+\lambda_1 H_{st}+H_F$, with λ_1 varying from 1 to 0. As $\lambda_1 = 0$, H_0+H_F possesses the approximate quantum number:

$$\sum |C_{i_s j_t}|^2 |i_s - j_t|$$

For convenience, we denote it as $|n_s - n_t|$. We have this approximate constant of motion because $(a_s^+ a_b a_b + a_t^+ a_b a_b + h.c.)$ increases or decreases the actions of s and t simultaneously in a form of linear combination. By the diabatic correlation, we can label $|n_s - n_t|$ on the eigenstates of the full Hamiltonian. They are the approximate constants. (The effect of H_{st} is smaller in $H_0+H_{st}+H_F$)

From $|n_s - n_t|$ and n_b $(=2(P-(n_s+n_t)))$, we can calculate (n_s, n_t). They are the approximately conserved quantities of the eigenstates of $H_0+H_{st}+H_F$. Table15.3 shows $|n_s - n_t|$, n_s+n_t and (n_s, n_t) of the 36 levels of D_2O with $P = 7$. Listed in the Table are also their formal quantum numbers $[n_s, n_t]$. Obviously, the truncated (n_s, n_t) are identical to $[n_s, n_t]$.

Table 15.3 The approximately conserved $|n_s - n_t|$, n_s+n_t and (n_s, n_t) of the 36 levels of D_2O with $P = 7$ and their formal quantum numbers $[n_s, n_t]$

Level	$\|n_s-n_t\|$	n_s+n_t	(n_s, n_t)	$[n_s, n_t]$
1	0	0	(0, 0)	[0 0]
2	0.98	1	(0.99, 0.04)	[1 0]
3	1.04	1	(1.02, −0.02)	[1 0]
4	2.00	2	(2.00, 0.00)	[2 0]
5	2.01	2	(2.00, −0.01)	[2 0]
6	0	2	(1.00, 1.00)	[1 1]
7-8	3.02	3	(3.01, −0.01)	[3 0]
9	0.96	3	(1.98, 1.02)	[2 1]
10-11	4.03	4	(4.02, −0.02)	[4 0]
12	1.05	3	(2.03, 0.98)	[2 1]
13	2.01	4	(3.01, 1.00)	[3 1]
14-15	5.05	5	(5.03, −0.03)	[5 0]
16	1.99	4	(3.00,1.01)	[3 1]
17-18	6.33	6	(6.17, −0.17)	[6 0]

19-20	6.5	7	(6.75,0.25)	[7 0]
21	3.00	5	(4.00,1.00)	[4 1]
22	0	4	(2.00,2.00)	[2 2]
23	3.00	5	(4.00,1.00)	[4 1]
24-25	3.99	6	(5.00,1.01)	[5 1]
26	0.97	5	(2.99,2.02)	[3 2]
27-28	4.96	7	(5.98,1.02)	[6 1]
29	1.99	6	(4.00,2.00)	[4 2]
30	1.03	5	(3.02,1.99)	[3 2]
31	2.99	7	(5.00,2.01)	[5 2]
32	1.98	6	(3.99,2.01)	[4 2]
33	2.99	7	(5.00,2.01)	[5 2]
34	0	6	(3.00,3.00)	[3 3]
35	0.99	7	(4.00,3.01)	[4 3]
36	1	7	(4.00,3.00)	[4 3]

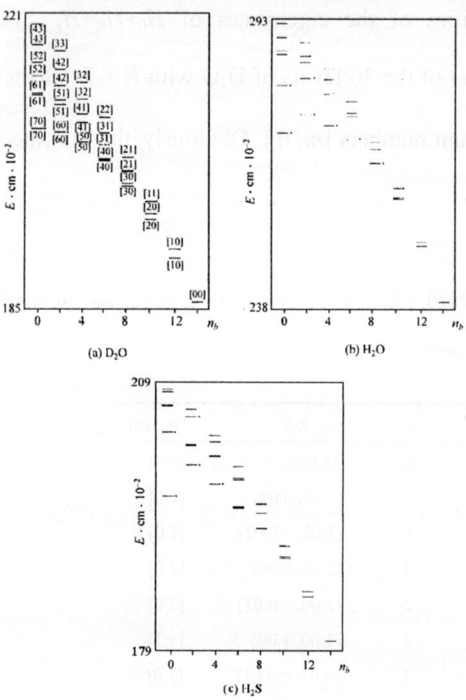

Fig.15.6 (a) The reconstruction of the 36 levels of D_2O with $P = 7$ by $[n_s, n_t]$.

(b),(c) are those of H_2O and H_2S. · shows that energy spacing is small.

Fig.15.6(a) shows the reconstruction of the 36 levels of D_2O with $P = 7$ by $[n_s, n_t]$. Fig.15.6(b) and(c) show the reconstructions of H_2O and H_2S. From these, we conclude:

(1) The energy levels of H_2O, D_2O, H_2S are different. However, from the reconstructed energy diagrams, they share the same *pattern*.

(2) For each n_b, there is a series of states with $[n_s, n_t]$. Lower levels have different n_s and n_t. Higher levels have closer n_s and n_t. The energy spacings are determined mainly by K_{st}. For the three molecules, the spacing is largest for D_2O, followed by H_2O and H_2S. This is consistent with the ordering of their K_{st}.

(3) For each series of levels with a common n_b, lower levels possess less energy spacings and are almost degenerate on some occasions. Then, the coupling between *s* and *t* stretches is weak and their motions are quite independent. For higher levels, the coupling is stronger and the character of normal mode shows up. Then, the energy spacings are nearly equi-distant. This is the character of a simple harmonic oscillator.

(4) Since approximate quantum numbers are retrieved from $|n_s - n_t|$, we cannot distinguish two levels with (n_s, n_t) and (n_t, n_s). Also, the two eigenstates of H_0 corresponding to $[n_s, n_t]$ and $[n_t, n_s]$ are degenerate and indistinguishable. Hence, in each series of levels with a common n_b, there are always two levels with the same $[n_s, n_t]$ labeling. This is a shortcoming. This can be remedied by labeling the two states with '+', '-'. '+' labelling corresponds to the symmetric motion of the *s* and *t* stretches and is with higher energy. '-'labelling corresponds to the antisymmetric motion and is with lower energy. Of course, '+', '-' are exact symmetries since the algebraic Hamiltonian is symmetric with respect to the interchanging of subscripts *s* and *t*. In other words, '+', '-' as operators commute with the Hamiltonian.

15.6 DCN case

The bend motion of DCN can be separated from the D–C (denoted as s) and C–N (denoted as *t*) stretches. The resonance forms between the stretches are 1:1 and 2:3 (This was discussed in Chapter 14.) The Hamiltonian can be written as:

$$H = H_0 + H_{st} + H_K$$

$$H_0 = \omega_s(n_s + \frac{1}{2}) + \omega_t(n_t + \frac{1}{2}) + X_{ss}(n_s + \frac{1}{2})^2 + X_{tt}(n_t + \frac{1}{2})^2$$
$$+ X_{st}(n_s + \frac{1}{2})(n_t + \frac{1}{2})$$

$$H_{st} = K_{st}(a_s^+ a_t + h.c.)$$

$$H_K = K(a_s^+ a_s^+ a_t a_t a_t + h.c.)$$

The coefficients are listed in Table 14.1.

H_0 will not destroy n_s, n_t. For H_{st}, $P_1 = n_s + n_t$ is conserved. For H_K, $P_2 = n_s/2 + n_t/3$ is conserved. For H, except energy, there are no constants of motion. This is a *simple* case but with very *complicated* dynamics. For its dynamical details, see Chapter 14.

By switching off H_K and the diabatic correlation, we have the approximate constant of motion P_1:

$$P_1 = \sum (i_s + j_t) | C_{i_s j_t} |^2$$

Similarly, if H_{st} is switched off, we have another approximate constant P_2:

$$P_2 = \sum (i_s / 2 + j_t / 3) | C_{i_s j_t} |^2$$

From P_1, P_2, (n_s, n_t) can be obtained. For the 66 levels of DCN, they are tabulated in Table15.4. Quite surprisingly, all (n_s, n_t) are integers. This shows that they are not seriously destroyed. (n_s, n_t) are exactly the formal quantum numbers in this case. We can reconstruct these 66 levels by P_1 as shown in Fig.15.7. The right most column is the 66 levels. No regularity is obvious. While for each P_1, the levels are labeled as $(n_s, n_t) = (0, P_1), (1, P_1-1), \ldots, (P_1, 0)$, consecutively. They show almost equal energy spacings. (With careful scrutiny, the spacings for the levels in the middle of a column are smaller. This corresponds to the separatrix. For details, see Chapter 9.) Therefore, by approximate quantum numbers, the levels, originally disordered, can be reconstructed into a very regular pattern.

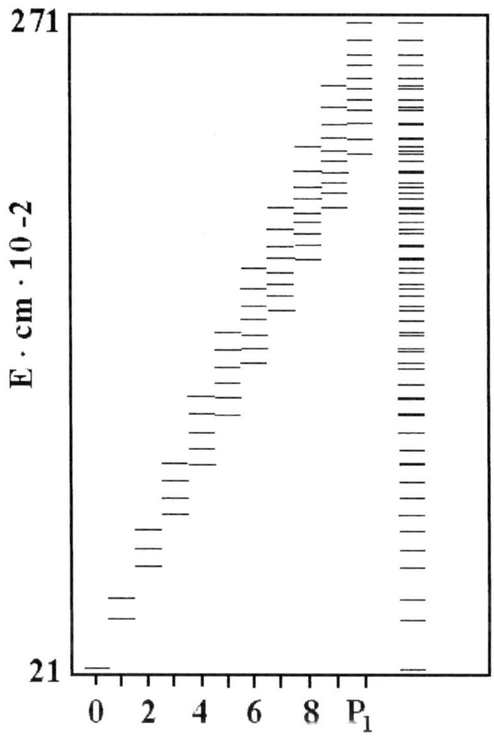

Fig.15.7 The reconstruction of the levels of DCN by P_1. For each P_1, the levels are labeled as $(n_s, n_t) = (0, P_1), (1, P_1\text{-}1),\ldots, (P_1, 0)$. The column on the far right has the 66 levels with no obvious regularity.

In conclusion, from the cases of formaldehyde, acetylene, H_2O, D_2O, H_2S and DCN, it is known that due to resonances, most conserved quantities are destroyed in the highly excited vibration. However, there are approximate constants of motion by which levels can be reconstructed into a very regular pattern. From the reconstructed diagram, dynamics concerning resonances and the origin of approximate quantum numbers become transparent.

Molecular highly excited levels, though immense and complicated, are assignable and can be classified.

Table 15.4 The approximate $P_1 = n_s + n_t$, $P_2 = n_s/2 + n_t/3$ and (n_s, n_t) for the 66 levels of DCN

Level	(P_1, P_2)	(n_s, n_t)
1	(0, 0)	(0, 0)
2	(1, 0.333)	(0,1)
3	(1, 0.5)	(1, 0)
4	(2, 0.666)	(0, 2)
5	(2, 0.833)	(1, 1)
6	(2, 0.999)	(2,0)
7	(3, 0.999)	(0, 3)
8	(3, 1.166)	(1, 2)
9	(3, 1.333)	(2,1)
10	(4, 1.333)	(0, 4)
11	(3, 1.5)	(3, 0)
12	(4, 1.5)	(1, 3)
13	(4, 1.666)	(2, 2)
14	(5, 1.666)	(0, 5)
15	(4, 1.833)	(3, 1)
16	(5, 1.833)	(1, 4)
17	(4, 2)	(4, 0)
18	(5, 2)	(2, 3)
19	(5, 2.166)	(3, 2)
20	(6, 2)	(0, 6)
21	(5, 2.333)	(4, 1)
22	(6, 2.166)	(1, 5)
23	(6, 2.333)	(2,4)
24	(5, 2.5)	(5, 0)
25	(6, 2.5)	(3,3)
26	(7, 2.333)	(0, 7)
27	(6, 2.666)	(4, 2)
28	(7, 2.499)	(1, 6)
29	(6, 2.833)	(5, 1)
30	(7, 2.666)	(2, 5)
31	(7, 2.833)	(3, 4)
32	(6, 3)	(6,0)
33	(7, 3)	(4, 3)
34	(8, 2.666)	(0, 8)
35	(7, 3.166)	(5,2)
36	(8, 2.833)	(1,7)
37	(8, 2.999)	(2, 6)
38	(7, 3.333)	(6, 1)
39	(8, 3.166)	(3,5)
40	(8, 3.333)	(4, 4)
41	(7, 3.5)	(7, 0)
42	(9, 2.999)	(0,9)
43	(8, 3.499)	(5, 3)
44	(9, 3.166)	(1, 8)
45	(8, 3.666)	(6,2)
46	(9, 3.333)	(2, 7)
47	(9, 3.5)	(3, 6)
48	(8, 3.833)	(7,1)
49	(9, 3.666)	(4, 5)
50	(10, 3.333)	(0,10)
51	(9, 3.833)	(5, 4)
52	(8, 4)	(8,0)
53	(10, 3.5)	(1,9)
54	(9, 4)	(6, 3)
55	(9, 4.166)	(7, 2)
56	(10, 3.666)	(2, 8)
57	(10, 3.833)	(3, 7)
58	(9, 4.333)	(8,1)
59	(10, 4)	(4, 6)
60	(10, 4.166)	(5, 5)
61	(9, 4.5)	(9, 0)
62	(10, 4.333)	(6, 4)

63	(10, 4.5)	(7, 3)
64	(10, 4.666)	(8, 2)
65	(10, 4.833)	(9, 1)
66	(10, 5)	(10, 0)

15.7 Difference between approximate and formal quantum numbers

The previous analysis of the cases of H_2O, D_2O, H_2S and DCN shows that, after truncation, the approximate quantum numbers are equal to the formal quantum numbers. However, this is not always the case. Table 15.5 shows the case of H_2O with $P = 9$. It is seen that after truncation, most approximate quantum numbers (n_s, n_t) are equal to the formal quantum numbers, while there are exceptions labeled by'*'. These situations are more complicated. The destruction of the constants of motion is more serious in these cases. However, the destruction is not too serious since we can still employ them to reconstruct the levels into a regular pattern as shown in Fig.15.8 which shows similar characters as the case with $P = 7$ in Fig.15.6(b).

Table 15.5 The approximate $|n_s\text{-}n_t|$, n_b, (n_s, n_t) and formal [n_s, n_t] for H_2O with $P = 9$.

level	$\|n_s\text{-}n_t\|$	n_b	(n_s, n_t)	[n_s, n_t]
1	0.07	18	(0.04,−0.04)	[0, 0]
2	1.08	16	(1.04,−0.04)	[1, 0]
3	0.95	16	(0.98,0.02)	[1, 0]
4	2.01	14	(2.01,−0.01)	[2, 0]
5	2.04	14	(2.02,−0.02)	[2, 0]
6	0.22	14	(1.11,0.89)	[1, 1]
7−8	3.08	12	(3.04,−0.04)	[3, 0]
9	0.92	12	(1.96,1.04)	[2, 1]
10	1.15	12	(2.08,0.92)	[2, 1]
11−12	4.23	10	(4.12,−0.12)	[4, 0]
13−14	8.96	0	(8.98,0.02)	[9, 0]
15	6.50	10	(5.25,−1.25)	[5, 0]*
16	6.50	8	(5.75,−0.75)	[5, 0]*
17	2.24	8	(3.62,1.38)	[3, 1]*
18	3.31	2	(5.66,2.34)	[8, 0]*
19	6.38	2	(7.19,0.81)	[8, 0]*
20	5.38	10	(4.69,−0.69)	[3, 1]*
21	0.34	10	(2.17,1.83)	[2, 2]
22−23	6.27	6	(6.14,−0.14)	[6, 0]
24−25	6.07	4	(6.54,0.46)	[7, 0]
26−27	3.18	8	(4.09,0.91)	[4, 1]
28−29	6.88	0	(7.94,1.06)	[8, 1]

30	1.74	8	(3.37,1.63)	[3, 2]
31	3.93	6	(5.35,0.65)	[5, 1]
32	4.69	6	(5.34,0.66)	[5, 1]
33	1.27	2	(4.64,3.36)	[3, 2]*
34	5.12	2	(6.56,1.44)	[7, 1]
35	5.12	8	(5.06,−0.06)	[7, 1]*
36−37	4.78	4	(5.89,1.11)	[6, 1]
38	2.04	6	(4.02,1.98)	[4, 2]
39	2.16	6	(4.08,1.92)	[4, 2]
40	0.44	4	(3.72,3.28)	[5, 2]*
41	4.39	0	(6.70,2.30)	[7, 2]
42	4.39	0	(6.70,2.30)	[7, 2]
43	3.73	6	(4.87,1.13)	[3, 3]*
44	3.73	4	(5.37,1.63)	[5, 2]
45	3.48	2	(5.74,2.26)	[6, 2]
46	3.51	2	(5.76,2.24)	[6, 2]
47	1.00	4	(4.00,3.00)	[4, 3]
48	2.36	2	(5.18,2.82)	[5, 3]
49	1.27	4	(4.14,2.86)	[4, 3]
50	2.26	0	(5.63,3.37)	[6, 3]
51	2.55	0	(5.78,3.22)	[6, 3]
52	2.30	2	(5.15,2.85)	[5, 3]
53	0.40	0	(4.70,4.30)	[4, 4]*
54	1.02	2	(4.51,3.49)	[5, 4]
55	0.89	0	(4.95,4.05)	[5, 4]

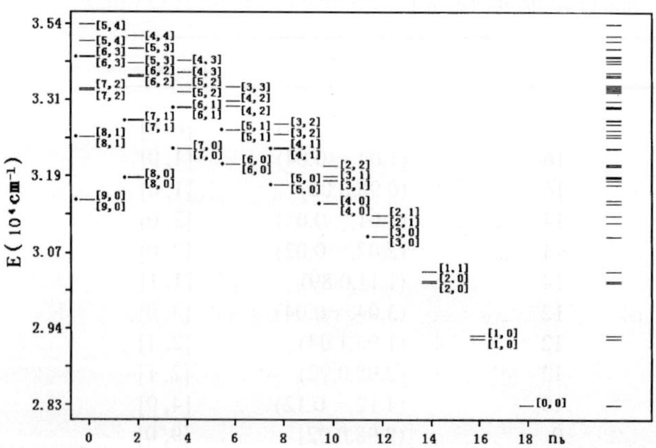

Fig.15.8 The reconstructed levels of H_2O with P=9. · shows that energy spacing is small.

15.8 Density ρ in the coset space

The Hamiltonians in the above cases can be cast to the expressions in the coset space. For H_2O, D_2O, and H_2S the coset space is SU(3)/U(2), for DCN it is two coupled H(4)/U(1)×U(1). Approximately, this algorithm can be realized via

Heisenberg's correspondence. Then, we have H expressed as $H\,(q_s, p_s, q_t, p_t)$. (q_s, p_s, q_t, p_t) are the coordinates of the coset spaces. (For details, see Chapters 2,7,14.)

Given an eigenenergy E, from $E = H(q_s, p_s, q_t, p_t)$, we can calculate (q_s, p_s, q_t, p_t). As long as the number of solutions is large, we can calculate the density ρ in the phase space.

An eigenstate is a linear combination of $\{|i_s, j_t\rangle\}$. This is the concept of quantum mechanics. In the phase space, the dynamical property is realized by ρ. For the quantal $\langle i_s, j_t\rangle$, there are the corresponding (q_s, p_s, q_t, p_t) points. The correlation of multiple (q_s, p_s, q_t, p_t) points to a given $\langle i_s, j_t\rangle$ is obvious. Of course, there are some $\langle i_s, j_t\rangle$ for which no (q_s, p_s, q_t, p_t) points can be found. For brevity, the density $\rho(\langle i_s, j_t\rangle)$ corresponding to $\langle i_s, j_t\rangle$ is shortened to $\rho(i_s, j_t)$.

The algorithm for calculating $\rho(i_s, j_t)$ is to mesh the whole space into 1.6×10^9 grids (q_α, p_α coordinates are partitioned into 200 grids, respectively.) .For the given i_s, j_t, we calculate the solutions satisfying the following conditions : (with H_2O as an example):

$$|\,H(q_\alpha, p_\alpha) - E\,| < 0.1\,(\mathrm{cm}^{-1})$$

$$|\,(q_s^{\,2} + p_s^{\,2})/2 - i_s\,| < 0.1$$

$$|\,(q_t^{\,2} + p_t^{\,2})/2 - j_t\,| < 0.1$$

The number of the solutions can then be assigned as $\rho(i_s, j_t)$. For the formal quantum numbers $[n_s, n_t]$, the corresponding $\rho([n_s, n_t])$ can also be calculated.

The calculation shows that as P is small, the correlation between $\rho(i_s, j_t)$ and $|C_{i_s j_t}|^2$ is not good. As P is large, the correlation becomes good. Shown in Fig.15.9 is the case of the 110th level (denoted as L110) of H_2O, $P = 15$. The correlation is apparent. Fig.15.10 shows the case of L40 of DCN. The correlation between $\rho(i_s, j_t)$ and $|C_{i_s j_t}|^2$ is also apparent. We note that ρ is a classical quantity, $|C_{i_s j_t}|^2$ is the concept of quantum mechanics. They are calculated independently. Their consistency is just the Correspondence Principle! In these two Figures, $\rho([n_s, n_t])$ are also shown. It seems that it is a rule that $\rho([n_s, n_t])$ is always larger.

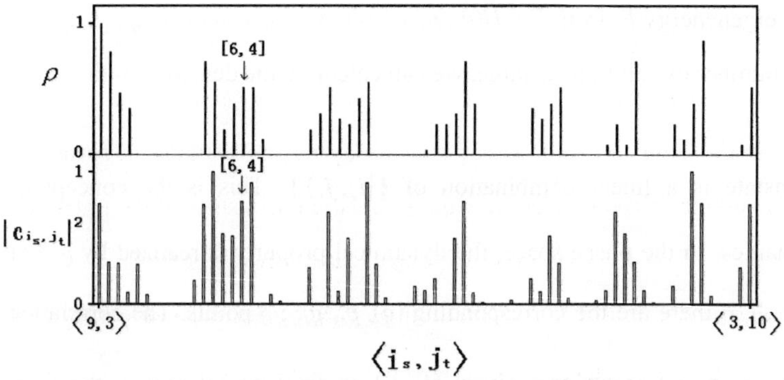

Fig.15.9 The correlation between $\rho(i_s, j_t)$ and $|C_{i_s j_t}|^2$ for the 110th level of H_2O, $P = 15$.

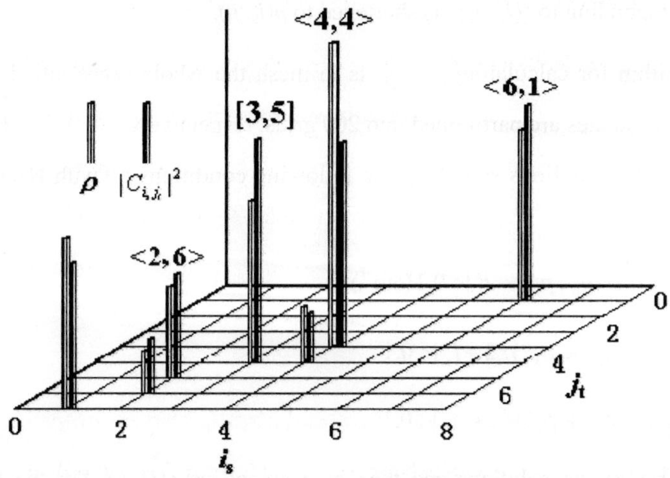

Fig.15.10 The correlation between $\rho(i_s, j_t)$ and $|C_{i_s j_t}|^2$ for the 40th level of DCN, $P = 10$.

15.9 Lyapunov exponents

From a solution $(q_s, p_s, q_t, p_t)_0$, as time t elapses a trajectory $(q_s, p_s, q_t, p_t)_t$ forms according to the equations of motion:

$$\partial H(q,p)\big/\partial p_\alpha = dq_\alpha\big/dt, \quad \partial H(q,p)\big/\partial q_\alpha = -dp_\alpha\big/dt \qquad (\alpha = \text{s, t})$$

Starting from a nearby point, there is another trajectory. Suppose the initial deviation between these two trajectories is δ_0, then as time elapses, the deviation δ_t can be:

$$\delta_t = \delta_0 e^{\lambda t}$$

If $\lambda > 0$, the two trajectories will diverge while if $\lambda < 0$, they will converge. λ is the Lyapunov exponent. It is an important dynamical parameter. In general, λ depends on the initial point. In some situations, it may be initial-point independent. The number of Lyapunov exponents is the same as the dimensionality of the trajectory. For a Hamiltonian system, Lyapunov exponents are always paired. For our four-dimensional space, due to the conservation of energy, the trajectories are in a three-dimensional subspace. Hence, the Lyapunov exponents are λ_{max}, 0, $-\lambda_{max}$ and only λ_{max} needs calculation (For details, see Chapter 13.).

In general, the calculation of λ_{max} is by the following method.

For an initial point x(0), a nearby point x′(0) with a distance of $|\, x(0) - x'(0)| = d_0$ is chosen. Note that both x(0) and x′(0) must lie on the same energy surface. Suppose after a duration of τ, the distance between the two evolving points, x(τ) and x′(τ) is d_1. Then a new point x″(τ) which lies in between x(τ) and x′(τ) with a distance of d_0 from x(τ) is chosen. By repeating the procedure, a series of d_i (i=1, …, n) can be obtained. Then the maximal Lyapunov exponent λ_{max} is: $\lambda_{max} = \lim_{n\to\infty} k_n$ with $k_n = \sum \ln (d_i/d_0) / n\tau$.

In the numerical calculation, we set k_n as λ_{max} when k_n reaches a constant value for large n. In another case, k_n shows a behavior of the form $\alpha n^{-\beta}$ ($\alpha, \beta > 0$). Then as n approaches infinity, k_n approaches zero and λ_{max} is zero.

In our calculation, τ is 0.53 fs, n is about 2–6×10^5, d_0 is 10^{-6} (as q_α, p_α are normalized to 1). Corresponding to each $\langle i_s, j_t \rangle$, there are many $(q_s, p_s, q_t, p_t)_0$ and, hence, there are as many λ_{max}. Fortunately, they are very close to each other.

The calculation shows that λ_{max} corresponding to the formal quantum number $[n_s, n_t]$ is always smaller than those of other $\langle i_s, j_t \rangle$. This is consistent with the fact that $[n_s, n_t]$ are the approximate constants. Table 15.6 shows the results of the case of H_2O, $P = 9$, L24. Apparently, λ_{max} of [7, 0] is the smallest. Listed in Table15.7 are the results for the case of DCN L20, L25, L30. λ_{max} corresponding to the formal quantum numbers are always the smallest.

Table 15.6 The Lyapunov exponents for H_2O, $P = 9$, L24

$\langle i_s, j_t \rangle$	λ_{max} (ps^{-1})
[7,0]	0.10
$\langle 5,0 \rangle$	0.11
$\langle 6,0 \rangle$	0.13
$\langle 3,2 \rangle$	0.15
$\langle 2,2 \rangle$	0.17
$\langle 8,1 \rangle$	0.19
$\langle 6,1 \rangle$	10.05
$\langle 4,1 \rangle$	10.21
$\langle 5,1 \rangle$	10.28
$\langle 3,1 \rangle$	10.86

Table 15.7 The Lyapunov exponents for DCN, L20, L25, L30

level	$\langle i_s, j_t \rangle$	λ_{max} (ps^{-1})
	[0,6]	0.04
20	$\langle 2,3 \rangle$	3.01
	$\langle 3,2 \rangle$	4.91
	$\langle 4,1 \rangle$	3.57
	[3,3]	0.06
25	$\langle 4,2 \rangle$	4.38
	$\langle 2,4 \rangle$	7.92
	[2,5]	0.59
	$\langle 3,4 \rangle$	11.40
30	$\langle 4,2 \rangle$	13.21
	$\langle 5,1 \rangle$	8.99
	$\langle 1,6 \rangle$	9.83
	$\langle 4,3 \rangle$	9.60

To show that the points in the phase space corresponding to the formal quantum number possess the least Lyapunov exponent, or that their trajectories are more regular, the Poincare surface (q_s, p_s) (p_t<0, along q_t= 0) structures of the trajectories originating from [3, 3] and $\langle 4,2 \rangle$ for the case of DCN, L25 are shown in Fig. 15.11(a), (b). Indeed, the trajectory originating from [3, 3] is quite regular while that from $\langle 4,2 \rangle$ is chaotic.

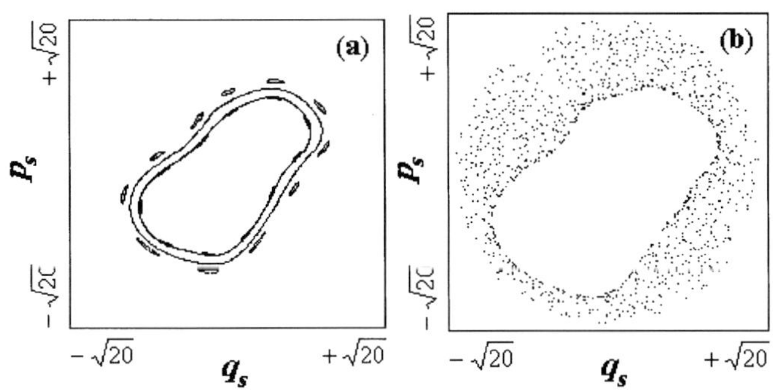

Fig.15.11 The Poincare surface (q_s, p_s) (p_t<0, along q_t = 0) structures of the trajectories originating from [3, 3] (a) and $\langle 4,2 \rangle$ (b) for the case of DCN, L25

References

15. 1 S. K. Gray and M. J. Davis, *J. Chem. Phys.*, 90 (1989) 5420.

15. 2 D. E. Reisner, R. W. Field, J. L. Kinsey and H. L. Dai, *J. Chem. Phys.*, 80 (1984) 5968.

15. 3 S. A. B. Solina, J. P. O'Brien, R. W. Field and W. F. Polik, *J. Phys. Chem.*, 100 (1996) 7797.

15. 4 M. E. Kellman and G. Chen, *J. Chem. Phys*. 95 (1991) 8671.

15. 5 Z. Lu and M. E. Kellman, *Chem. Phys. Lett*. 247 (1995) 195.

15. 6 J. P. Rose and M. E. Kellman, *J. Chem. Phys*. 105 (1996) 7348.

15. 7 G. Wu, *Chem. Phys. Lett*., 292(1998)369.

15. 8 Z. Ji and G. Wu, *Chem. Phys. Lett*., 311(1999)467.

15. 9 D. Zheng and G. Wu, *Acta Physica Sinica* 19(2002)66.

15. 10 P. Wang and G. Wu, *Chem. Phys. Lett*., 371(2003)238.

Chapter 16

One-electronic motion in multiple sites

16.1 Classical analogues of LCAO of one-electronic system

For the electron molecular orbital theory (MO), the system wavefunction ψ can be the linear combination of the atomic (or hydrogen-like) orbitals (LCAO), φ_i:

$$\psi = \sum C_i \varphi_i$$

Here, C_i is the linear combination coefficient. $|C_i|^2$ shows the probability that ψ is found on φ_i. The sign of C_i shows the phase relation in ψ. If C_i and C_j of two atomic orbitals (wavefunctions), φ_i and φ_j, are of the same sign, then φ_i and φ_j are *in resonance* and between them is the bonding orbital. Conversely, if C_i and C_j are of opposite signs, then φ_i and φ_j are *out of resonance* and antibonding forms. If there is no interaction between φ_i and φ_j, then there is nonbonding and the corresponding C_j (C_i) is 0. LCAO is the core concept of quantum mechanics. If we interpret $|C_i|^2$ as the action and the sign relation between C_i, C_j as the angle, that is if C_i, C_j are of the same sign, the angle difference is 0; if they are of opposite signs, the angle difference is π; for nonbonding, the angle difference is $\pi/2$. In this way, we have the classical analogues, action and angle, for the MO coefficients.

The analogy of MO coefficients with action/angle variables as interpreted above will be helpful in its association with the coset representation. This is essential for our treatment of one-electronic dynamics in the coset framework.

16.2 Hamiltonian of one electron in multiple sites: the coset representation

The Hamiltonian of one electron in multiple sites can be written as:

$$\sum e_i a_i^+ a_i + 1/2 \sum V_{kl} a_k^+ a_l$$

Here, e_i is the site energy at site *i*, $a_i^+ a_i$ is the population of electrons or action and $a_k^+ a_l$ is the interaction between sites *l* and *k*. $V_{kl}/2$ is its interaction energy.

In Section 2.4, we have addressed the dynamical space of one electron in *n* sites being the *n* dimensional hypersphere S^n. In this coset representation, the Hamiltonian (for its details, see Section 2.5) is the sum of the following terms: First, choose an arbitrary site and label it as site 0. The remaining sites are labeled 1 to $n-1$.

(1) The energy at these sites is

$$e_0(1 - \sum_{i=1}^{n-1}(q_i^2 + p_i^2)) + \sum_{i=1}^{n-1} e_i(q_i^2 + p_i^2)$$

(2) The interaction energy among the sites is the sum of

$$\sum_{i=1}^{n-1} V_{0i} q_i (1 - \sum_{i=1}^{n-1}(q_i^2 + p_i^2))^{1/2}$$

and

$$\sum_{i \neq j \neq 0} V_{ij}(q_i q_j + p_i p_j)$$

Here, $n_i = q_i^2 + p_i^2 \leq 1$ shows the action at site *i*. $1 - \sum q_i^2 + p_i^2$ is the action at site 0. $\theta_i = \tan^{-1}(-p_i/q_i)$ shows the phase angle between the actions at sites *i* and 0. $q(= q_1, q_2, \ldots q_{n-1})$, $p(= p_1, p_2, \ldots p_{n-1})$ are the coordinates of S^n.

Hamilton's equations of motion are:

$$\partial H(q,p)/\partial p_i = dq_i/dt$$

$$\partial H(q,p)/\partial q_i = -dp_i/dt$$

Starting from an initial point $(q,p)_0$, by the integration of the equations, we then have a trajectory in the coset phase space.

16.3 Analogy with Hückel MO

The above-mentioned one-electron Hamiltonian is consistent with the Hückel molecular orbital (HMO) method. e_i and $V_{ij}/2$ correspond to α and β parameters in HMO. (See Ref.16.1 for HMO). To demonstrate this, we will discuss the system of three identical linear sites with nearest neighboring interaction. (HMO also considers the nearest neighboring interaction.) The parameters are $e_0 = e_1 = e_2 = -8.1 \times 10^4\,\text{cm}^{-1}$, $V_{01} = V_{12} = -1.4 \times 10^4\,\text{cm}^{-1}$, $V_{02} = 0$。

$H(q,p)$ can give the classical energy range. The quantized energies by HMO are in the classical energy range. We partition the energy range into 10 intervals and calculate their (q,p) and thereby (n_0, n_1, n_2), respectively, as shown in Fig.16.1. The three HMO quantized energies are in the 1st, 5th, and 10th intervals. For them, (n_0, n_1, n_2) are (0.25, 0.5, 0.25),(0.5, 0, 0.5) and (0.25, 0.5, 0.25), respectively. The eigencoefficients by HMO are (0.5, 0.707, 0.5),(0.707, 0, -0.707) and (0.5, -0.707, 0.5), respectively. Obviously, n_i is roughly proportional to $|C_i|^2$.

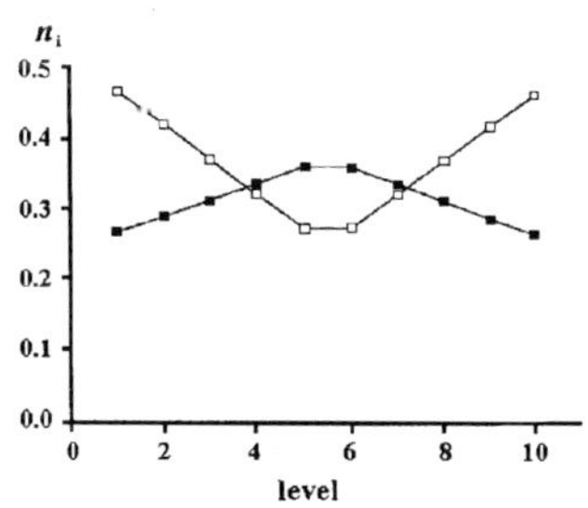

Fig.16.1 The classical energy range is partitioned into 10 intervals and the calculated ■(n_0, n_2), □(n_1). See text for details.

We can calculate θ_1 and θ_2 for the classical levels from (q,p). For the 1st energy range, both θ_1 and θ_2 are 0. This is consistent with the signs of the eigencoefficients of the first HMO level — they are all positive. For the 5th energy range, the distribution of θ_1 and θ_2 is shown in Fig.16.2 in which θ_1 and θ_2 are concentrated around $\pi/2$ and π, respectively. That θ_1 is around $\pi/2$ is consistent with $C_1 = 0$ of the 2nd level of HMO. That θ_2 is concentrated around π indicates that the actions on sites 0 and 2 are out of phase. Indeed, C_0 and C_2 are of opposite signs.

Fig.16.2 The distribution of θ_1 (heavy line) and θ_2 (light line) for the 5th level

For the 10th energy range, the distribution of its θ_1 and θ_2 is shown in Fig.16.3 in which θ_1 and θ_2 are concentrated around π and 0, respectively. This means that C_1 and C_2 should be negative and positive, respectively. Indeed, this is consistent with the eigencoefficients of the 3rd quantum level by HMO.

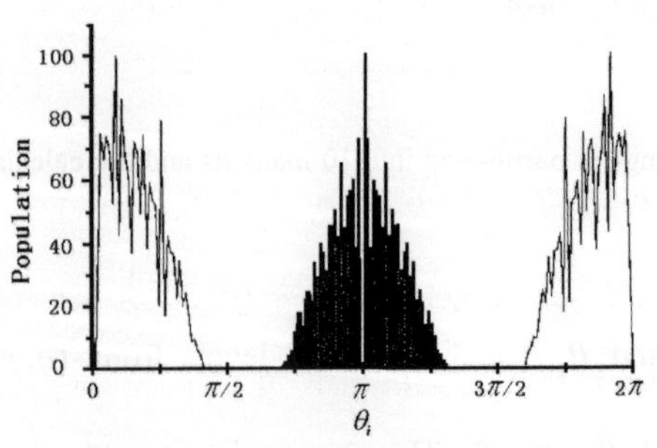

Fig.16.3 The distribution of θ_1 (heavy line) and θ_2 (light line) for the 10th level

The above analysis shows that action and angle defined by the coset representation are in congruence with the square of the magnitude and sign of the eigencoefficients by quantum mechanics, respectively.

16.4 Dynamical interpretation of HMO

By the assertion mentioned previously, that there is correspondence between (n_i,θ_i) and C_i and its sign, we can obtain $(n^0{}_i,\theta_i^0)$ from the eigencoefficients by HMO. Then, if we use them as the initial conditions to calculate the evolving trajectory, this will offer a dynamical interpretation for HMO. As an example, in the following we will consider two systems of linear and ring structures of three identical sites. Shown in Table.16.1 are their three levels and the associated C_i (also its sign) and $(n^0{}_i,\theta_i^0)$.

Table16.1 Eigencoefficients C_i and $(n^0{}_i,\theta_i^0)$ for the HMO levels (denoted by L#) of linear and ring structure of three identical sites.

		(C_0, C_1, C_2)	$(n^0{}_1,\theta_1^0)$	$(n^0{}_2,\theta_2^0)$
Linear sites	L1	(0.5, 0.71, 0.5)	(0.5, 0)	(0.25, 0)
	L2	(0.71, 0, -0.71)	(0, $\pi/2$)	(0.5, π)
	L3	(0.5, -0.71, 0.5)	(0.5, π)	(0.25, 0)
Ring sites	L1	(0.58, 0.58, 0.58)	(0.336, 0)	(0.336, 0)
	L2	(0.71, -0.71, 0)	(0.5, π)	(0, $\pi/2$)
	L3	(0.41, 0.41, -0.82)	(0.168, 0)	(0.672, π)

Calculations show that all these trajectories are periodic and θ_i is concentrated around θ_i^0. Besides, for the linear structure, the trajectories for the highest and lowest levels have the tendency to mingle with each other as shown in Fig.16.4.

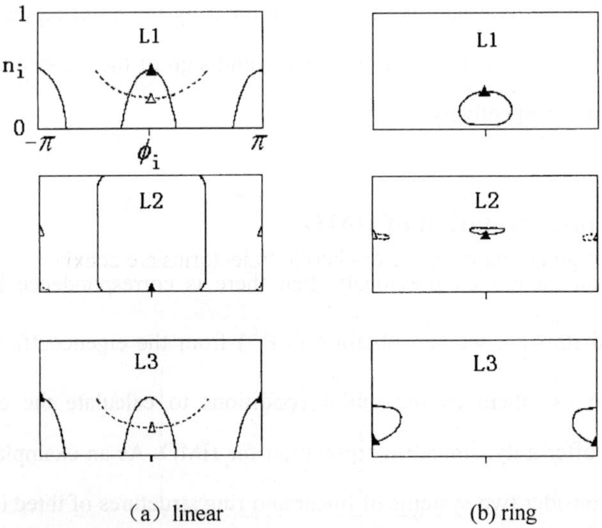

Fig.16.4 $n^0{}_1, \theta_1{}^0$ (▲) and its evolution (——) $n^0{}_2, \theta_2{}^0$ (△) and its evolution (…)

In (b) L1, L3 (▲), (——) and (△), (……) coincide with each other.

As mentioned, the coset space of one electron in three sites is S^3. For these SU(3) systems, dynamical chaos is possible. The trajectories of HMO levels are special in that they are periodic. In fact, chaotic trajectories are possible for the asymmetric interactions. Shown in Fig.16.5 are the chaotic trajectories of the lowest level for which $n_1^0 = 0.247$, $\theta_1{}^0 = 0$, $n_2^0 = 0.352$, $\theta_2{}^0 = 0$ with asymmetric interactions: V_{02}:V_{01}:V_{12}=10:7:5. The interesting point is that though the trajectories are chaotic, they are still around the initial points.

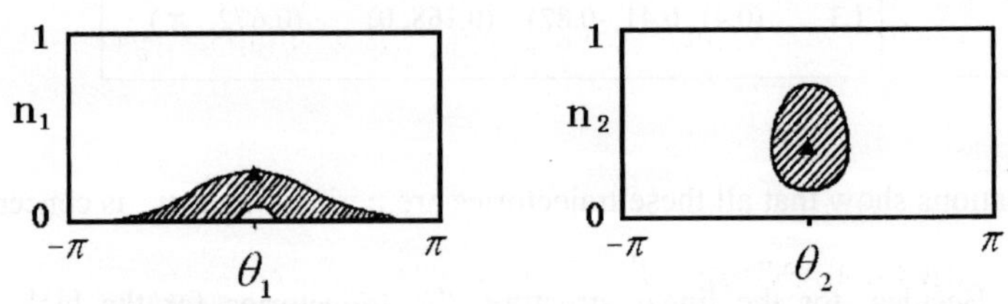

Fig.16.5 The chaotic trajectories of the lowest level with asymmetric interactions.
▲ is the Huckel point, shaded regions are the chaotic regions.

This picture can be extended to the motion of two non-interacting electrons.

Fig.16.6 shows the Poincare surface of section along (q_2, p_2) ($q_1 = 0$, $p_1 > 0$) with $n_1^0 = 0.958$, $n_2^0 = 0.604$ as two electrons are in the two lowest levels (no spin space is considered). The parameters are: $V_{01} = -2100$ cm^{-1}, $V_{02} = -3000$ cm^{-1}, $V_{12} = -1500$ cm^{-1} and $e_i = 0$. In this case, θ_i^0 cannot be uniquely defined and multiple values were chosen. Obviously, regular and irregular or chaotic trajectories are coexistent.

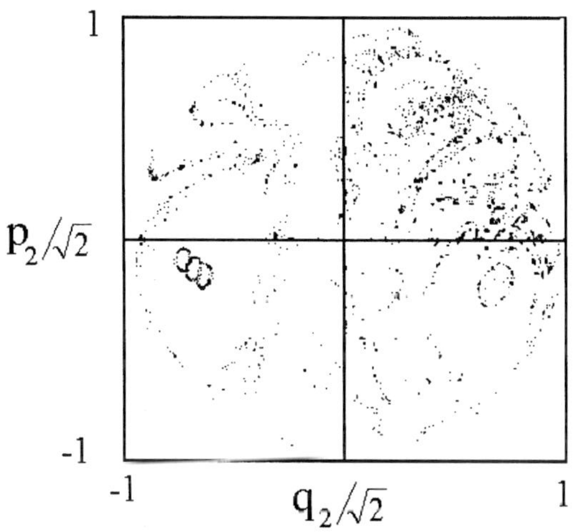

Fig.16.6 The Poincare surface of section along (q_2, p_2) ($q_1 = 0$, $p_1 > 0$) with $n_1^0 = 0.958$, $n_2^0 = 0.604$ as two electrons are in the two lowest levels. See text for the details.

We have tried to interpret HMO by the coset concept. As demonstrated, the HMO states correspond to the special points in the multi-dimensional coset space. For asymmetric interactions, chaotic trajectories prevail. Then, an important issue which awaits solution is how to characterize the dynamics from a global sense in the coset space!?

16.5 Anderson localization

There is potential application of our approach to the one-electronic system. Anderson localization and the Hammett equation for organic reactions are two such cases that will be addressed.

Anderson localization indicates that in a random lattice under certain conditions, the electron diffusion will be hindered. Anderson has pointed out that for the three-dimensional lattice with random distribution in site energy in a range of W, and with the nearest neighboring (the number is Z) interaction V, the electronic motion will be localized in a small region if $W/(2Z|V|) > e$ (e is the transcendental number.). For a one-dimensional system, electronic motion is always localized regardless of V. For the case of a two-dimensional system, localization is not conclusive though it is generally believed that electronic motion is localized. (Ref.16.2-4)

For modeling the localization, we consider $n_0 = 1$, i.e., q_i and p_i are 0. The systems are cubic, square and linear lattices. The nearest neighboring interaction energy V is -1.4×10^4 cm^{-1}. The site numbers are, respectively, 10^3, 10^2 and 20. The site energy e_i is randomly chosen. e_0 is -2×10^4 cm^{-1}.

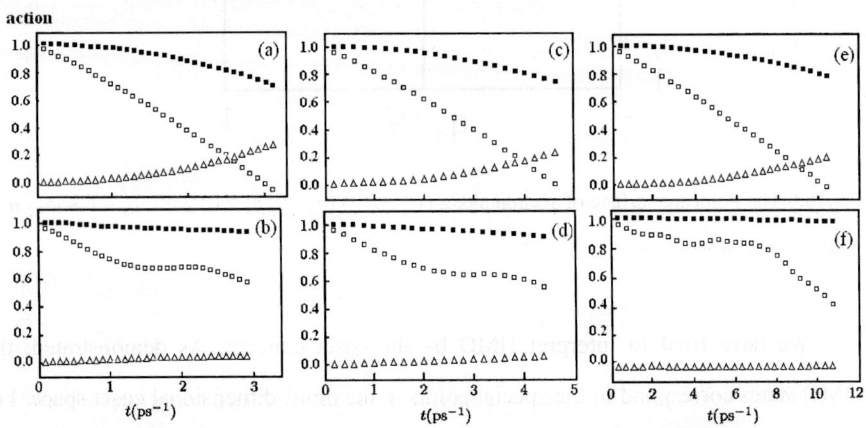

Fig.16.7 The action evolution of n_0 (□), $n_0+\sum n_i$ (■) (i is the nearest neighbor of site 0) and $\sum n_j$ (□) (j is the rest site) for 3-dimension (a), (b); 2-dimension (c), (d); and 1-dimension (e), (f). (a), (c), (e) with $W = 0$; (b), (d), (f) with $W = 100\times10^4$ cm^{-1}

Fig.16.7 shows the result with W = 0 and 100×10^4 cm^{-1} (Larger W is taken for better visualizing the result) of n_0 (□), $n_0+\sum n_i$ (■) (i shows the nearest neighbors of site 0) and $\sum n_j$ (△)(j shows the rest sites) as a function of time. Very obviously, random distribution of e_i will lead to the relaxation retardation of $n_0+\sum n_i$. The time scale in the Figure is not important since different orders of V, e_i will lead to different time scales. At a specific time, the ratio of $\sum n_j/Z$ with W = 100×10^4 cm^{-1} to that with W = 0 is the smallest for the one-dimensional system. This is consistent with the general recognition that localization is most obvious in the one-dimensional system.

In this numerical modeling, though it is hard to have a definitive and sharp definition for localization, the basic characteristic of Anderson localization is vividly expressed.

16.6 Hammett equation

In organic reactions, we are interested in the effect of substitutes on the reaction rate. As shown in Fig.16.8, reaction takes place at Y and different substitutes at X will have different impacts on the reaction rate.

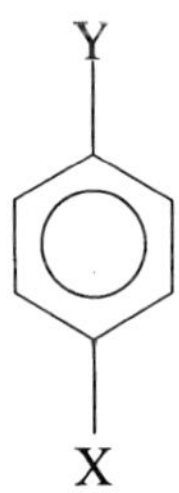

Fig.16.8 Reaction takes place at Y and different substitutes at X will have different impacts on the reaction rate.

Hammett has summarized that there can be a linear relationship between log(reaction rate k) and the character parameter σ for a series of substitutes:

$$\log k = \sigma\rho + a$$

Here, ρ and a are constants. σ is the character parameter for the substitute X. Hammett's equation is especially useful for the chemical reactions that are independent of steric effect.

For modeling, we consider the nearest neighboring interaction with V = -1.4×10^4

cm^{-1}. The interaction of a substitute with its nearest neighbor is V_X. We imitate σ with V_X. Meanwhile, all e_i is 0. At time $t = 0$, suppose an electron is at X. The *transfer rate* of the electron from X to Y is considered as the effect of X on the reaction rate at Y where the reaction takes place. Suppose t_c is the time required for the action at Y to reach n_Y^c which is a given parameter. The reaction rate is thus defined as $\log(n_Y^c / t_c)$. In the literature, the range of σ for various substitutes is around 2.5 (Ref.16.5). From $\Delta G = -RT \ln k$, we estimate the range of ΔG to be of the order of 10^3 cm^{-1}. Meanwhile, we suppose the range of V_X is about 10^3 cm^{-1} which is much smaller than V. For validating the linearity of Hammett's equation, we expand the range of V_X ten fold.

Fig.16.9 shows the results for n_Y^c= 0.01(a), 0.1(b) and 0.5(c). The linearity of ($\log(n_Y^c / t_c)$) against V_X is very obvious.

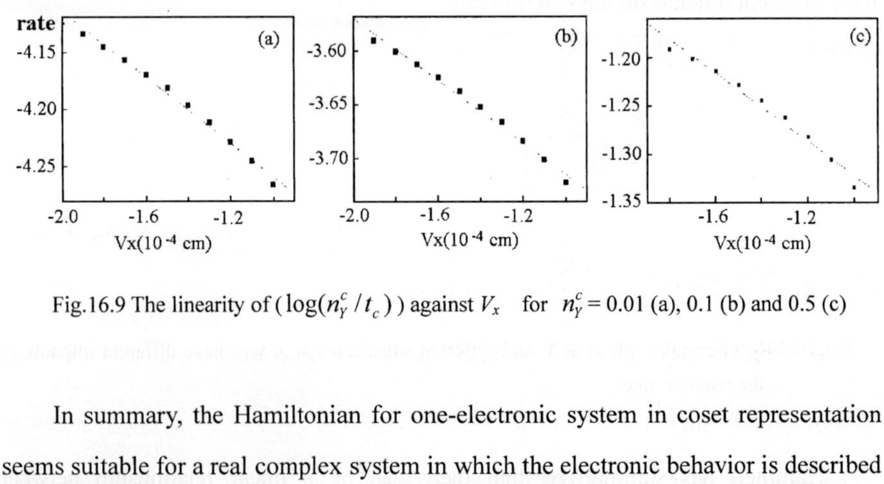

Fig.16.9 The linearity of ($\log(n_Y^c / t_c)$) against V_X for n_Y^c = 0.01 (a), 0.1 (b) and 0.5 (c)

In summary, the Hamiltonian for one-electronic system in coset representation seems suitable for a real complex system in which the electronic behavior is described by the trajectories in the coset space. The calculation is straightforward. This is different from the traditional quantal method, like the molecular orbital theory which involves the solution of wavefunctions. From the viewpoint of dynamics, this algorithm seems superior since wavefunction is not so convenient for the description of dynamics.

16.7 Two-electronic correlation in Hückel system

Electron correlation is an important topic in MO. As a model, the simplest case is two electrons in three orbitals. In Section 2.5, we have addressed the suitability of coset representation for this issue. Here, we consider HMO as an example.

Before demonstrating this, we first summarize the coset results by SO(6)/U(3). The system Hamiltonian is :

$$2(1+\frac{3}{2}\sum(q_i^2+p_i^2))^{-1}$$
$$\times[(q_1^2+p_1^2)(e_2+e_3)+(q_2^2+p_2^2)(e_1+e_3)+(q_3^2+p_3^2)(e_1+e_2)]$$
$$+2[1+\frac{3}{2}\sum(q_i^2+p_i^2)]^{-1}$$
$$\times[V_{12}(q_1q_2+p_1p_2)+V_{23}(q_2q_3+p_2p_3)-V_{13}(q_1q_3+p_1p_3)]$$

Here, e_i(i = 1, 2, 3) is the energy at the i orbital. $\sum(q_i^2+p_i^2)<2$. The number of paired electrons are

$$4\sum(q_i^2+p_i^2)\,[1+\frac{3}{2}\sum(q_i^2+p_i^2)]^{-1},$$

and

$$n_i=q_i^2+p_i^2$$

shows the paired electron numbers in orbitals j and $k(\neq i)$. (Strictly speaking, a factor of $2[1+\frac{3}{2}\sum(q_i^2+p_i^2)]^{-1}$ has to be multiplied.)

$\phi_i=\tan^{-1}(-p_i/q_i)$ shows the phase angle. As time elapses, if ϕ_i only takes certain specific values, then the correlation is stable. Otherwise, if ϕ_i is extensive in a certain range, then the correlation stability is poor.

This coset formulation offers us a classical model for electronic correlation.

Table16.2 shows the pairing patterns and the phase angles derived from the Hückel eigencoefficients (See Table 16.1) of the lincar and ring lattices of three sites. Phase 0 and π indicate that the signs of the coefficients of the orbitals are of the same sign or not, respectively. For instance, for the 1st level of the linear lattice, the actions on the three sites are (0.25, 0.5, 0.25). Hence, there are two pairing configurations: （0, 0.25, 0.25） and (0.25, 0.25, 0). The phase angles are 0.

Table16.2 The pairing patterns and the phase angles derived from the Hückel eigencoefficients (See Table 16.1) of the linear and ring lattices of three sites

	level	paring configuration $(s_1, s_2, s_3; \phi)$	evolving angles $(\phi_1; \phi_2; \phi_3)$
Linear lattice	L1	(0, 0.25, 0.25; 0)	$(0,\ \pi;\ \pm\pi/2; 0,\ \pi)$
		(0.25, 0.25, 0; 0)	
	L2	$(0.5, 0, 0.5; \pi)$	$(\pm\pi/2; 0,\ \pi; \pm\pi/2)$
	L3	$(0, 0.25, 0.25; \pi)$	$(0,\ \pi;\ \pm\pi/2; 0,\ \pi)$
		$(0.25, 0.25, 0; \pi)$	
Ring lattice	L1	(0, 0.1665, 0.1665; 0)	quasi-periodic in $(-\pi, \pi)$
		(0.1665, 0, 0.1665; 0)	
		(0.1665, 0.1665, 0; 0)	
	L2	$(0, 0.5, 0.5; \pi)$	$(\approx\pm\pi/3;\ \approx 0,\ \pm 2\pi/3; \approx\pm\pi/3, \pi)$

Taking these pairing configurations and angles as the initial conditions and by Hamilton's equations, we can obtain the evolving actions and angles (n_i, ϕ_i). Parameters are $e_i = 0, V_{ij} = 1000$ cm^{-1}. Fig.16.10 shows that for a linear lattice, ϕ_i are strictly limited in certain values. This indicates that the electron pairing or correlation of the Hückel orbitals of a linear lattice is very stable. Table16.2 also shows these angles (ϕ_1, ϕ_2, ϕ_3). For the 1st level of ring lattice, (n_i, ϕ_i) shows quasi-periodic behavior while the 2nd level shows the angles are limited to certain ranges as shown in Fig.16.11. Linear and ring lattices are different in that in the ring lattice each site is two-sided connected while in the linear lattice, the end points are only one-sided connected. This difference results in that the electronic pairing in the ring lattice is more unstable. For the linear lattice, the pairing is with fixed angles.

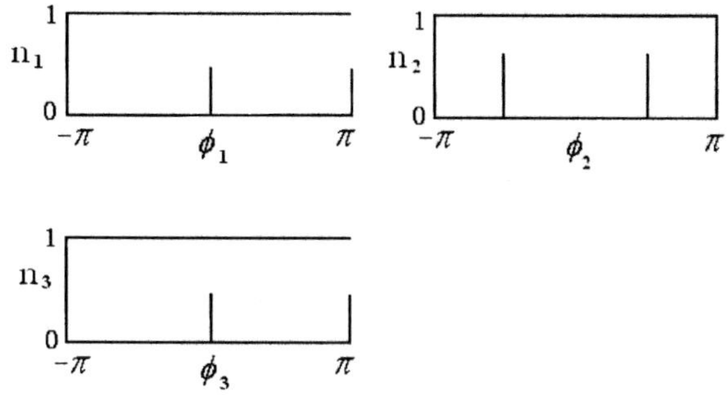

Fig.16.10 For the 1st level of linear lattice, ϕ_i are strictly limited in certain values.

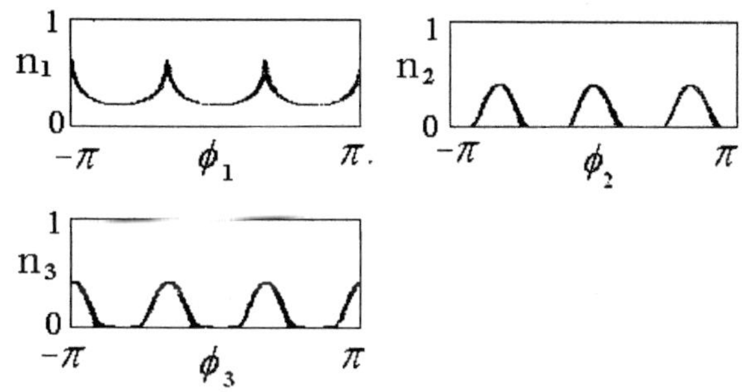

Fig.16.11 For the 2nd level of ring lattice, the angles are limited to certain ranges.

Calculation also shows that the increase of e_i will expand the range of ϕ_i, i.e., will not be prone to the pairing stability while the increase of V_{ij} will force ϕ_i to certain specific values, i.e., will enhance pairing or its stability. The unevenness of e_i and V_{ij} will destroy the pairing. This is consistent with the physical intuition. Besides, we have tried the non-Hückel orbitals. They just do not possess the pairing property.

In summary, from the viewpoint of coset dynamics, Hückel orbitals possess nontrivial properties.

References

16. 1 A. Streitwieser, *Molecular Orbital Theory for Organic Chemists*, Wiley, New York, 1961.

16. 2 P. W. Anderson, *Phys. Rev.*, 109(1958)492.

16. 3 E. Abrahams, P. W. Anderson, D. C. Licciardello and T. V. Ramarkrishnan, *Phys. Rev. Lett*. 42(1979)673.

16. 4 S. Sarker and E. Domany, *Phys. Rev.* 23B(1981)6018.

16. 5 J. March, *Advanced Organic Chemistry: Reactions, Mechanisms and Structure*, McGraw-Hill, New York, 1968.

16. 6 G. Wu, *Chem. Phys. Lett*., 343(2001)339.

16. 7 G. Wu, *Chem. Phys. Lett*., 246(1995)413.

16. 8 G. Wu, *Chem. Phys. Lett*., 264(1997)398.

16. 9 G. Xing and G. Wu, *Chem. Phys. Lett*. 18(2001)157.

Chapter 17

Lyapunov exponent, action integrals of periodic trajectories and quantization

17.1 Introduction

Bohr first pointed out the quantization in a Coulomb field. Sommerfeld then extended the idea to that in a system; if the space coordinate q is a cyclic variable then the integration of its momentum p in space will be a multiple of Planck's constant h. Later, Einstein, Brillouin and Keller (EBK) gave the quantization condition of an integrable system as:

$$\oint pdq = h(n+\delta)$$

The integral is called the action. δ is the Maslov index. n is an integer. For vibration and electronic motion, δ =1/2. For in-plane rotation, δ =0. However, the semiclassical quantization of a nonintegrable or chaotic system has not been solved. Recently, Gutzwiller has pointed out that under semiclassical approximation, the quantal density of states can be related to the periodic trajectories. Gutzwiller's formula involves the convergence of infinite series and the whole or complete exploitation of periodic trajectories is not an easy task. Indeed, there is the difficulty of the realistic application of his trace formula. (Ref.17.1)

Our work involves one-electronic motion in a lattice and molecular vibration like H_2O. In the coset representation, the system dynamics is (semi)classical and nonlinear. The dynamics is realized by the trajectories in the coset space. On the one hand, many concepts in nonlinear dynamics such as fractal, chaos and the Lyapunov exponent can find application in this field. On the other hand, the main shortcoming of this algorithm is its inability to provide quantized levels. The essence of the issue is the long-time unsolved problem: the semiclassical quantization of nonintegrable or chaotic systems.

In this chapter, we will address how, via the analysis of the Lyapunov exponent, this problem can be partially solved or that it sheds some light on the issue in certain cases. It should be emphasised that our choice of trajectories is random and our quantization is by the least average of Lyapunov exponents. This is different from the viewpoint of Gutzwiller, which involves periodic trajectories, as our trajectories are hardly periodic. The number of our randomly chosen trajectories is 100 or 200. In averaging the Lyapunov exponents, those trajectories with larger Lyapunov exponents will play a more significant role.

In this chapter, we will also consider the quantization of DCN vibration by the actions of its periodic trajectories (Section 14.2). For the two coupled DCN stretch modes, the intermode couplings will lead to nonintegrability, i.e., the periodic trajectories will be destroyed with the birth of new periodic trajectories and chaotic trajectories (Section 3.4 for KAM theorem). Then, the quantization by action as pointed out by Bohr, Sommerfeld or EBK meets with difficulty. However, for not very high excitation, there is a possibility that actions of the remnant periodic trajectories can be employed for quantization, especially of those lower levels. This is what we intend to explore.

17.2 Hamiltonian for one electron in multiple sites

In Chapter 16, we have formulated the Hamiltonian of one electron in multiple sites in analogy with HMO. Here, four-site systems are chosen for study as shown in Fig.17.1. They are labeled as S1, S2, S3.

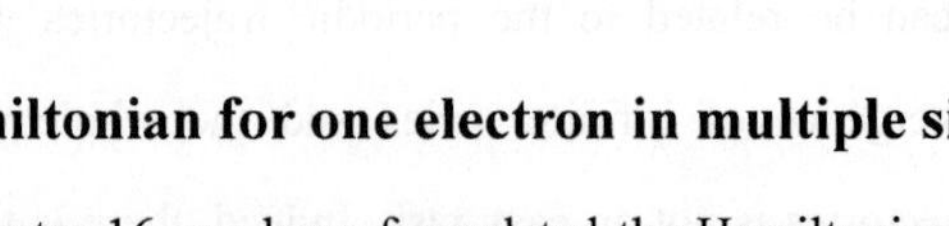

Fig.17.1 Systems of four-site lattices

The system Hamiltonian is

$$e_0[1-\sum_{i=1}^{3}(q_i^2+p_i^2)]+\sum_{i=1}^{3}e_i(q_i^2+p_i^2)$$

$$+\sum_{i=1}^{3}V_{0i}q_i[1-\sum(q_i^2+p_i^2)]^{1/2}+\sum_{i\neq j\neq 0}V_{ij}(q_iq_j+p_ip_j)$$

Here, e_0, e_1, e_2, e_3 are the site energies at sites 0, 1, 2, 3. e_i corresponds to the α value in HMO. V_{ij} is the interaction energy between sites *i* and *j*. $V_{ij}/2$ corresponds to β in HMO. The nearest neighbor approximation will be adopted, i.e., if *i*, *j* do not correspond to adjacent sites, then V_{ij}=0. In our calculation, we take e_i= -8.1×10^4 cm^{-1}, $V_{ij}=-1.4\times10^4$ cm^{-1}. Actions at the sites are: $n_0=1-\sum q_i^2+p_i^2$ and $n_i=q_i^2+p_i^2$ (i = 1, 2, 3). Since there is but one electron, we have $\sum q_i^2+p_i^2\leq 1$. Angle $\theta_i=\tan^{-1}(-p_i/q_i)$ shows the phase relation between actions n_0 and n_i. θ_i can be correlated with the signs of the linear combination coefficients, C_0 and C_i. If C_0 and C_i are of similar magnitudes and the same sign, then θ_i= 0. If they are of opposite signs, then $\theta_i=\pi$. If C_i=0, then $\theta_i=\pi/2$. Furthermore, if C_i>0, (suppose C_0>0) and its magnitude is quite small, then θ_i is smaller than but very close to $\pi/2$. Similarly, if C_i<0 with quite small magnitude, then θ_i is larger than but very close to $\pi/2$. $\sqrt{n_i}$ is proportional to $|C_i|$.

Let q_i, p_i run over all their allowed ranges, then the maximal and minimal $H(q,p)$ are exactly the highest and lowest level energies of HMO. $H(q,p)$ offers the continuous energy range as shown in Fig. 17.2(a). Our goal is to determine the quantized energies from $H(q,p)$.

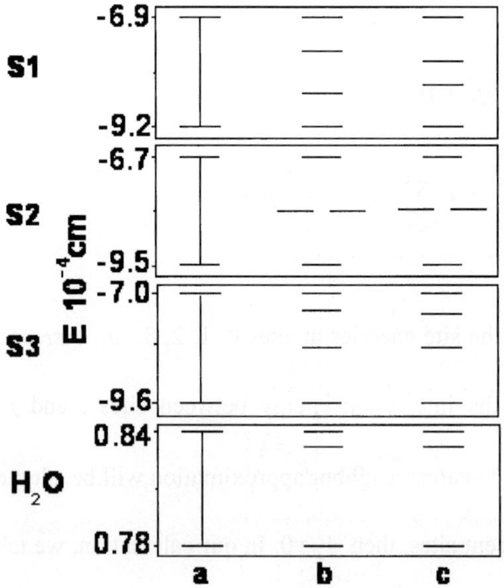

Fig.17.2 (a) classical energy range (b) quantized energies (c) energies with least $\langle\lambda\rangle$ for S1, S2, S3 and H_2O.

17.3 Quantization: the least of the averaged Lyapunov exponents

We partition $H(q,p)$ into 100 intervals. Corresponding to each energy at the center of an interval, there are many $(q_1, p_1, q_2, p_2, q_3, p_3)$ points. Thereof, we randomly choose 200 points and calculate their Lyapunov exponents. These exponents will span a range. Denote their average as $\langle\lambda\rangle$ and consider it as the average of that interval. (If we increase the partitions, the interval will be narrower.) Fig.17.3 shows the relation between $\langle\lambda\rangle$ and energy. Thereof, we have four local minima for S1 and S3. For S2, there are only three local minima. The locations of these minima are shown in Fig.17.2(c). The energies evidenced by local minimal $\langle\lambda\rangle$ are very close to the quantized energies by HMO (Fig.17.2(b)). For S2, due to symmetry, two quantized levels are degenerate. Indeed, in the three local minima of $\langle\lambda\rangle$, there is one that is correspondent to two degenerate levels.

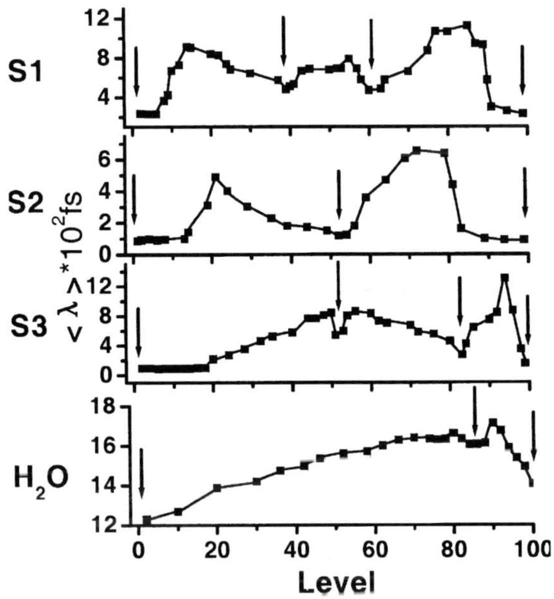

Fig.17.3 Local minima of averaged Lyapunov exponent $\langle\lambda\rangle$ for S1, S2, S3 and H_2O.

Hence, our conjecture is that quantized energy corresponds to the local minimum of averaged Lyapunov exponent.

This idea shares common ground with those of Bohr and Gutzwiller. In their ideas, only standing wave and periodic trajectories can be stable in a quantum system. Classically, a stable standing wave and periodic trajectories possess a zero Lyapunov exponent. However, this algorithm does not depend on periodic trajectories. Instead, it depends on the chaotic trajectories possessing larger Lyapunov exponents that contribute more to $\langle\lambda\rangle$. Hence, the quantization condition proposed is: least chaos in the global phase space.

One important observation is that the frequency spectra of the trajectories (that is the Fourier transform) of the levels with least $\langle\lambda\rangle$ are always simpler than the others.

Once the level energies with least $\langle\lambda\rangle$ are determined, the corresponding $(q_1,p_1,q_2,p_2,q_3,p_3)$ and $(\theta_1,\theta_2,\theta_3)$, (n_0,n_1,n_2,n_3) can be obtained. Numerically, they have a distribution and the maximal values of θ_i and n_i are what we will take to infer the signs of C_i and $\sqrt{n_i}$. These results are listed in Table17.1. The inferred signs of C_i are the same as those of HMO. Indeed, as C_i of HMO becomes smaller, the corresponding θ_i is closer to $\pi/2$ and the inferred C_i in the Table is 0. In the Table, though $\sqrt{n_i}$ and C_i of HMO are not exactly identical, their relative magnitudes are consistent. For the 3rd level of S3, the inferred sign of C_2 is negative and its magnitude is the largest, however, it is 0 by HMO. Another inconsistency is for this level $\sqrt{n_3}$ =0.44, while by HMO, C_3 is 0 and C_3 is also assigned as 0 due to $\theta_3 \sim \pi/2$.

Table 17.1 The inferred signs of C_i and $\sqrt{n_i}$ for S1, S2, S3 and H_2O. See text for details.

system	level	eigencoefficient (HMO) (C_0, C_1, C_2, C_3)	Coset algorithm (C_0, C_1, C_2, C_3) sign	Coset algorithm $(\sqrt{n_0}, \sqrt{n_1}, \sqrt{n_2}, \sqrt{n_3})$
S1	L1	(+0.37, +0.60, +0.60, +0.37)	(+, +, +, +)	(0.37, 0.60, 0.59, 0.37)
	L2	(+0.60, +0.37, -0.37, -0.60)	(+, +, 0, -)	(0.51, 0.48, 0.48, 0.51)
	L3	(+0.60, -0.37, -0.37, +0.60)	(+, -, -, +)	(0.51, 0.48, 0.48, 0.51)
	L4	(+0.37, -0.60, +0.60, -0.37)	(+, -, +, -)	(0.37, 0.59, 0.59, 0.37)
S2	L1	(+0.50, +0.50, +0.50, +0.50)	(+, +, +, +)	(0.50, 0.50, 0.49, 0.50)
	L2/L3	(+0.71, 0, -0.71, 0)	(+, 0, -, 0)	(0.50, 0.49, 0.49, 0.49)
		(0, +0.71, 0, -0.71)		
	L4	(+0.50, -0.50, +0.50, -0.50)	(+, -, +, -)	(0.49, 0.50, 0.49, 0.50)
S3	L1	(+0.52, +0.52, +0.61, +0.28)	(+, +, +, +)	(0.52, 0.52, 0.60, 0.28)
	L2	(+0.37, +0.37, -0.25, -0.82)	(+, +, 0, -)	(0.49, 0.48, 0.47, 0.53)
	L3	(+0.71, -0.71, 0, 0)	(+, -, -, ~0)	(0.51, 0.51, 0.52, 0.44)
	L4	(+0.30, +0.30, -0.75, +0.51)	(+, +, -, +)	(0.32, 0.32, 0.73, 0.50)
H_2O		$(C_{n_s}, C_{n_t}, C_{n_b})$	$(C_{n_s}, C_{n_t}, C_{n_b})$ sign	$(\sqrt{n_s}, \sqrt{n_t}, \sqrt{n_b})$

	L1	(+0.04, +0.04, +0.99)	(+, +, +)	(0.08, 0.08, 0.99)
	L2	(+0.71, +0.71, -0.06)	(+, +, 0)	(0.72, 0.57, 0.35)
	L3	(+0.71, -0.71, 0)	(+, -, 0)	(0.72, 0.64, 0.17)

17.4 Quantization of H_2O vibration

The algebraic vibrational Hamiltonian of H_2O, $H(q_t, p_t, q_b, p_b)$, can be written as:

$$\omega_s(n_s + n_t + 1) + \omega_b(n_b + \frac{1}{2})$$
$$+ X_{ss}[(n_s + \frac{1}{2})^2 + (n_t + \frac{1}{2})^2] + X_{bb}(n_b + \frac{1}{2})^2$$
$$+ X_{st}(n_s + \frac{1}{2})(n_t + \frac{1}{2}) + X_{sb}(n_s + n_t + 1)(n_b + \frac{1}{2})$$
$$+ K_{st}(2n_s)^{\frac{1}{2}} q_t + K_{sb}\{\sqrt{n_s}(q_b^2 - p_b^2) + [q_t(q_b^2 - p_b^2) + 2p_t q_b p_b]/\sqrt{2}\}$$

with

$$n_t = (q_t^2 + p_t^2)/2, \quad n_b = (q_b^2 + p_b^2)/2, \quad n_s = N - n_t - n_b/2$$

Here, s and t stand for the two O–H stretches and b for the HOH bend. ω and X are harmonic and anharmonic coefficients. K_{st}, K_{sb} are the coefficients for the 1:1 coupling between s and t and Fermi resonance among s, t and b, respectively. Their elucidation and values have been stated in detail in Chapter 7.

As an example, take $N = 1$. Fig.17.2 shows the energy range by $H(q_t, p_t, q_b, p_b)$ and the three quantum levels by the second quantized Hamiltonian (See Section 1.4 and Chapter 7 for details.). We partitioned the classical energy range into 100 intervals and randomly chose 200 initial points of the solution space corresponding to the energy at the center of each interval to calculate the Lyapunov exponents and their average $\langle\lambda\rangle$. The results are shown in Fig.17.3. Indeed, there are three local minima. These level energies by locally minimal Lyapunov exponents are also shown in Fig 17.2（c）. They are very close to the quantum energies (Fig.17.2(b))

From the energies corresponding to minimal $\langle\lambda\rangle$, we can obtain (q_t, p_t, q_b, p_b)

and θ_i, n_i from which the signs of C_i and magnitudes $\sqrt{n_i}$ can be elucidated. As shown in Table 17.1, the results are quite consistent with those by quantum mechanics.

17.5 A conjecture

The viewpoint that quantized energies correspond to locally minimal Lyapunov exponents is noted. The viewpoint concerns the minimal degree of chaoticity in the global phase space. In the calculation of $\langle\lambda\rangle$, initial points are randomly chosen. The issue then is: which trajectories play the crucial role in this algorithm of quantization and what causes the minimal degree of chaoticity in a global sense. Though our Hamiltonian is classical and of coset-represented form, this quantization viewpoint of a chaotic system may be prevalent and can be seen for general nonintegrable systems.

17.6 Action integrals of periodic trajectories

17.6.1 Periodic trajectories

In Section 14.2, we have analyzed the periodic trajectories for the coupled D–C and C–N nonlinear oscillators of DCN. Its detailed dynamical treatment has been treated in Section 14.1. For periodic trajectories, we can employ two Poincare surfaces of section for analysis: (q_s, p_s), q_t=0, p_t<0 and (q_t, p_t), q_s=0, p_s<0 and define the number of points left on the surface as their periods. For convenience,

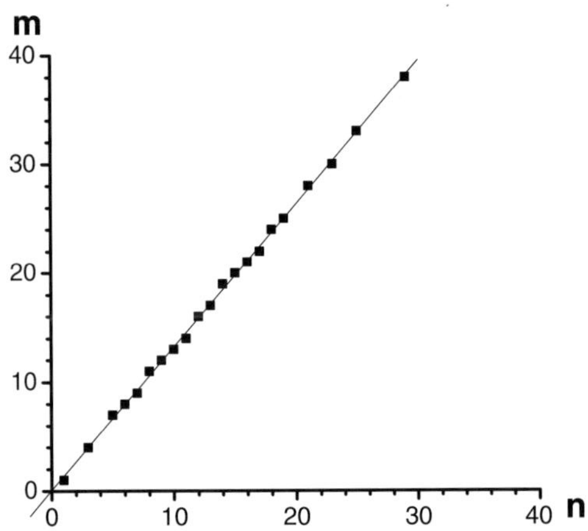

Fig.17.4 The relation between the periods n and m on the surfaces: (q_s, p_s) and (q_t, p_t) for periodic trajectories

we will use p^n or q^m to show that a periodic trajectory has n and m points on the surfaces of (q_s, p_s) and (q_t, p_t), respectively. For a periodic trajectory, there are two notions p^n and q^m. m is different from n, in general except for period-1 trajectories (see Section 14.2). For example, for p^3, (period-3 trajectory) we have q^4. For p^5, p^7, p^8, p^9, p^{12}, p^{15}, p^{18}, we have q^7, q^9, q^{11}, q^{12}, q^{16}, q^{20} and q^{24} (for p^5, we have also q^5). m and n show a nice linear relation as shown in Fig.17.4. The slope is 1.313. This value is very close to the harmonic frequency ratio of D–C and C–N : 1.3. In fact, this is the so-called winding number.

For an eigenstate, we may have several trajectories possessing the same p^n (q^m). For period-1 trajectories, we have two different p^1 (q^1) trajectories. Their actions are concentrated on either D–C or C–N and their displacements are, respectively, antisymmetric and symmetric. Hence, we will use p_a^1 (q_a^1) and p_s^1 (q_s^1) to represent them. Fig. 17.5 shows period-1 (L11) and period-3 (L22) trajectories and their intersections on the (q_s, p_s) and (q_t, p_t) surfaces.

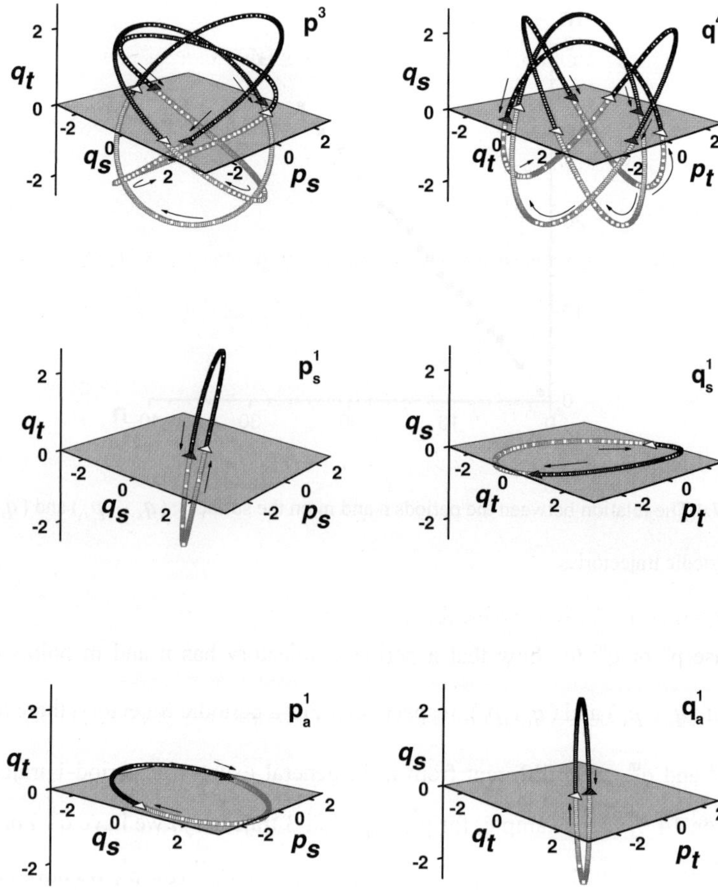

Fig.17.5 The period-1 (L11) and period-3 (L22) trajectories of DCN and their intersections on the (q_s, p_s) and (q_t, p_t) surfaces. ▲denotes p_s or p_t<0. △ denotes p_s or p_t>0. Arrows show the direction of the evolving trajectories.

17.6.2 Action integrals

For a periodic trajectory, its action integral, L, is defined as

$$L = \frac{1}{2\pi}\oint \vec{p}\cdot d\vec{q} = \frac{1}{2\pi}\left(\oint p_s dq_s + \oint p_t dq_t\right)$$

The integration is along the path of the trajectory. For the trajectories possessing the same p^n, their L are the same (very close to each other). Hence, we have L(p^n)

or $L(q^m)$. Similarly, we have $L(p_s^1)$ and $L(p_a^1)$. Fig.17.6 shows the relation of L (of various p^n) and E. Though the data are not complete, several conclusions can be drawn:

(1) For each p^n, (or p_s^1, p_a^1) the relation is linear. This is also true for q^m.

(2) For each E, L is linear against n and m. Fig.17.7 shows the situation for the 17th level (levels are counted from the lowest one). We also note that $L(p_s^1)$ and $L(p^n)$, $L(p_a^1)$ and $L(q^m)$ show a linear relation against period. Specifically, we have:

$$L(p^i) : L(p^j) \cong i:j$$

$$L(q^s) : L(q^t) \cong s:t$$

(3) For lower quantum states and smaller n, $L(p^n)$ ($L(q^m)$) is, or close to, an integer

(4) From Fig. 17.6, we have for the ground state:

$$L(p^n) = L(p_s^1) = L(p_a^1) = 0$$

The phase space of the ground state is only a point in which all the trajectories shrink to a point.

These rich linear behaviors enable us to calculate all $L(p^n)$ at any E from very few L values. For instance, suppose $L(p_s^1)$, $L(p_a^1)$ and $L(p^3)$ at any two E_1, E_2 are known, then from Fig. 17.6, $L(p_s^1)$, $L(p_a^1)$ and $L(p^3)$ at any E can be known. Meanwhile, for any E, since $L(p_s^1)$ and $L(p^3)$ are known, all $L(p^n)$ will be known just like the situation shown in Fig.17.7. Furthermore, from $L(p_a^1)$ and $L(q^4)$(= $L(p^3)$), all $L(q^m)$ will be known. (These $L(q^m)$ and $L(p^n)$ are interrelated.)

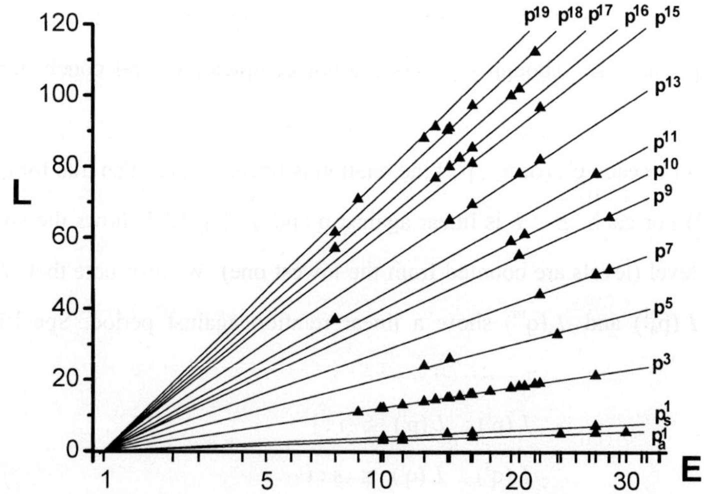

Fig.17.6 The dependence of action, L on E for various p^n.
The numbers on E axis show the eigenlevels.

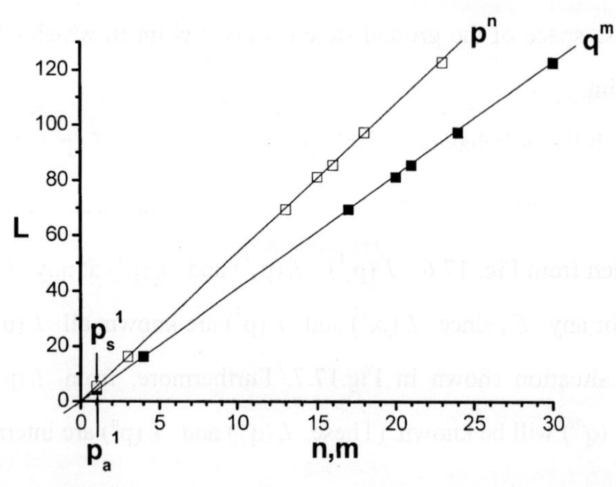

Fig.17.7 The dependence of $L(p^n)$, $L(q^m)$ on periods n and m for the 17th level of DCN

In the next section, we will show that these relations give us a simple and easy way to retrieve the low quantal levels.

17.7 Retrieval of low quantal levels

Fig.17.6 shows that quantal levels correspond exactly to integral $L\,(p_s^1)$, $L\,(p_a^1)$ or $L\,(p^3)$. The linear relations provide the following formulae for energy e_i as the functions of positive numbers (These can be nonintegers and are $L\,(p_s^1)$, $L\,(p_a^1)$, $L\,(p^3)$) :

$$e_1\,(\mathrm{cm}^{-1}) = 1919.1\,n_1 + 2304.4$$

$$e_2\,(\mathrm{cm}^{-1}) = 2521.7\,n_2 + 2453.4$$

$$e_3\,(\mathrm{cm}^{-1}) = 632.9\,n_3 + 2343.5$$

Here, n_1, n_2, n_3 are positive numbers.

We note that the period of a trajectory is reciprocal to the resolution of its energy spectrum. p_s^1, p_a^1 and p^3 offer the resolution of 1900 cm^{-1}, 2500 cm^{-1} and 600 cm^{-1}. These resolutions are enough to define the vibrational spectrum of DCN. Since p^3 offers finer resolution by p_s^1 and p_a^1, hence, n_3 will be defined by n_1 and n_2, i.e., the levels defined by n_3 should be in between the levels by n_1 and n_2.

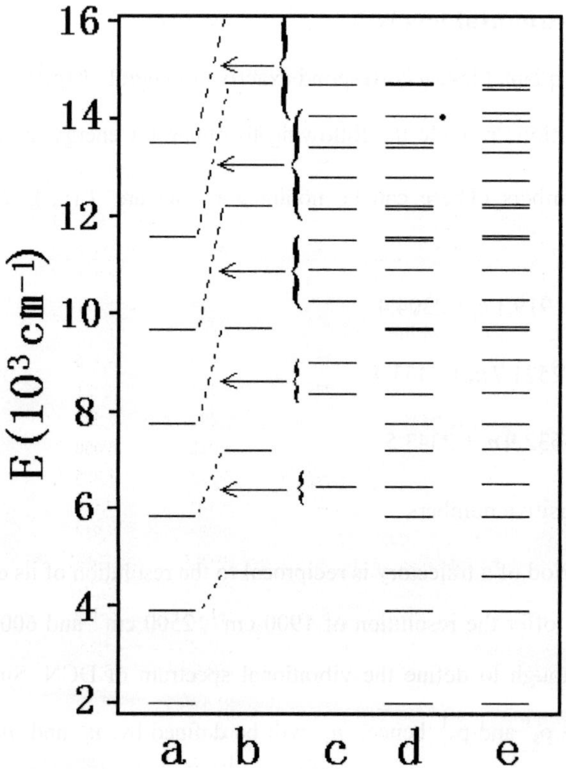

Fig.17.8 The quantal levels by integral (a) $L\,(p_s^{\,1})$, (b) $L\,(p_a^{\,1})$, (c) $L\,(p^3)$. Arrows and brackets show that in between the levels in (a), (b) connected by the broken lines, we can insert the levels with integral $L\,(p^3)$. (d) is the totality of the levels in (a), (b) and (c). In (e) are the quantal levels. '•' denotes the degenerate levels.

Quantal levels can be retrieved by the following procedure:

(i) Let $n_1 = 0, 1, 2, \ldots$, we have the levels as shown in Fig.17.8(a).

(ii) Let $n_2 = 1, 2, \ldots$, we have the levels as shown in Fig.17.8(b). (Note n_2=0 and n_1= 0 offer quite the same level)

(iii) Among the levels in Figure 17.8 (a) and (b) that are connected by the broken lines (corresponding to $n_1 = n_2 \equiv n_0$), insert proper e_3 levels (corresponding to $n_3 = 3n_0+1,\ 3n_0+2,\ \ldots,\ 4n_0-1$) as shown in Figure 17.8 (c). Figure 17.8 (d) is the totality of the levels in Figure 17.8 (a), (b) and (c). They are very close to the quantal levels in Figure 17.8 (e). Table17.2 shows these level energies. The deviation is not larger than 100 cm^{-1}, i.e., less than 1%. We have to stress that this method is suitable to low levels, less than 15000 cm^{-1} since there is difficulty in determining the periodic trajectories for high levels.

Table17.2 The levels by integral $L(p_s^1)$, $L(p_a^1)$, $L(p^3)$ and by the quantal method (cm^{-1})

level	by p_s^1	by p_a^1	by p^3	totality	quantal level
1	2,304			2,304	2,301
2	4,224			4,224	4,221
3		4,975		4,975	4,934
4	6,143			6,143	6,139
5			6,774	6,774	6,821
6		7,497		7,497	7,521
7	8,062			8,062	8,060
8			8,673	8,673	8,705
9			9,305	9,305	9,364
10	9,981			9,981	9,984
11		10,018		10,018	10,062
12			10,571	10,571	10,599
13			11,204	11,204	11,200
14			11,837	11,837	11,852
15	11,900			11,900	11,915
16			12,470	12,470	12,502
17		12,540		12,540	12,556
18			13,103	13,103	13,051
19			13,736	13,736	13,628
20	13,819			13,819	13,851
21			14,369	14,369	14,278
22			14,369	14,369	14,425
23			15,002	15,002	14,915
24		15,062		15,062	15,001

In summary, the whole procedure starts from two randomly chosen energies which are very probably not the eigenenergies. Determine their p^3, p_s^1 and p_a^1 trajectories and the actions. Follow the method shown in the last section to determine the linear dependence of $L(p_s^1)$, $L(p_a^1)$ and $L(p^3)$ on energy. Then, from the linear functions, quantized energies can be elucidated.

For DCN, 1:1 and 2:3 couplings between D–C and C–N stretches destroy the otherwise conserved quantum numbers such as n_s+n_t, $n_s/2+n_t/3$. However, the destruction is not so serious that they are still approximately conserved. (See Chapter 15). An interesting point is that $L(p_s^1)$ and $L(p_a^1)$ of the levels in Fig.17.8(a) and (b) are just the approximate constants of motion: $n_s + n_t$ (They are polyad number P_1 in Section 15.6). This implies the close relation between $L(p^n)$ and (approximately) conserved quantum number. Their exact relation can bear important significance for the systems of highly excited vibration.

17.8 Conclusion

For a nonintegrable system, due to the destruction of periodic trajectories, its quantization via action integrals was suspected. Einstein was the first to point out this difficulty for a nonintegrable system. Our work shows that for low levels, the situation may not be so pessimistic. The remnant periodic trajectories in a chaotic system possess quantities that are related to quantization. In this aspect, Gutzwiller's idea about periodic trajectories is crucial. On the other hand, we have the conjecture that quantization can be based on the locally minimal Lyapunov exponents. That is: quantization is related to the minimal degree of chaoticity in a global sense. The implication is that chaotic and periodic trajectories are intrinsically related in a nonintegrable system. Though the KAM theorem demonstrates the evolution of trajectories under perturbation, there is still very much we do not know about for future exploration.

References

17. 1 M. C. Gutzwiller, *Chaos in Classical and Quantum Mechanics in Interdisciplinary Applied Mathematics*, Springer, Berlin,1990.

17. 2 P. Wang and G. Wu, *Phys. Rev. A*, 66, 022116(2002).

17. 3 P. Wang and G. Wu, *Chem. Phys. Lett.* 375(2003)279.

Chapter 18

Application of the *H* function in vibrational relaxation

18.1 The *H* function

Classical dynamics is symmetric with respect to time reversal. Newtonian function $f = ma$ or Hamilton's equations are symmetric under $t \to -t$ transformation. However, our world seems irreversible. Everything always moves toward the future. It is never in reverse. In physics, the quantity for irreversibility is entropy, *S*. Entropy is a macroscopic idea. How to relate micro-reversible process to the macro-irreversible process, or how to derive macro-irreversible phenomena from a micro-reversible process is a fundamental physical problem. Boltzmann has derived a distribution function for a many-particle system to construct the so-called *H* function which decreases as time elapses till the system reaches equilibrium. In other words, the *H* function is the foundation for entropy. The relation is:

$$H = -S/k_B$$

Here, k_B is Boltzmann's constant. The significance of Boltzmann's work is the derivation of irreversible process from time-reversal dynamics.

In the following, in analogy with the *H* function, we will construct a function that increases with time in the coset dynamics. This analogous *H* function can be used to characterize the relaxation of molecular vibration.

Intramolecular vibrational energy redistribution, (IVR) plays an important role in chemical dynamics. Traditional IVR is based on RRKM (Rice-Ramsperger-Kassel-Marcus) theory. This theory presumes that intramolecular excitation will relax to an equilibrium state in a very short time scale. Though it can explain many phenomena of relaxation, there are also many observations that relaxation does not follow the way that RRKM theory predicts. For instance, many IVR processes are not ergodic.

18.2 Construction of the *H* function for vibrational relaxation

We will construct an analogous *H* function for IVR, taking the vibrational systems of H_2O and DCN as examples. The coset Hamiltonians for these two systems have already been described in Chapter 7 and 14.

Given initial $(n_\alpha, n_\beta)_0\ (\alpha = s,t)$, the corresponding points in phase space are $(q_\alpha, p_\alpha, q_\beta, p_\beta)_0$. Initially, suppose as $(n_\alpha, n_\beta) = (n_\alpha, n_\beta)_0$, the density of states $\rho(n_\alpha, n_\beta, t = 0) = 1$ (ρ is normalized) and as $(n_\alpha, n_\beta) \neq (n_\alpha, n_\beta)_0$, $\rho = 0$. As time elapses, by Hamilton's equations, $\rho(n_\alpha, n_\beta, t)$ can be determined. Define the *H* function as:

$$H(t) = -\sum_{n_\alpha}\Big[\sum_{n_\beta}\rho(n_\alpha, n_\beta, t)\ln\sum_{n_\beta}\rho(n_\alpha, n_\beta, t)\Big]$$

$H(t)$, thus defined, will increase as time elapses and finally reach a constant. (For convenience, a negative sign is added to ensure that *H(t)* increases always.)

The points $(q_\alpha, p_\alpha, q_\beta, p_\beta)_0$ corresponding to the initial $(n_\alpha, n_\beta)_0$ span an energy interval. In our calculation, the energy chosen is close to the center of the energy interval and the deviation is less than 100 cm^{-1} (Alternatively, we may not limit the energy range. The choice of 100 cm^{-1} deviation is based on the consideration that a laser pulse of 1 ps width corresponds to the energy uncertainty of 100 cm^{-1}). In our calculation, we randomly chose 1000 $(q_\alpha, p_\alpha, q_\beta, p_\beta)_0$ points satisfying the energy condition. Then we calculate $\rho(n_\alpha, n_\beta, t)$ as a function of time. The step of integration is 0.53 fs and the total duration is 50 fs.

18.3 Resonances in H_2O and DCN vibration

Fig.18.1 is the evolving $H(t)$ as a function of time. Initially, $(n_s, n_t)_0 = (10,10)$. The increase of $H(t)$ as time elapses is obvious. As time is large, $H(t)$ approaches a constant. This is similar to entropy that in a spontaneous process will reach a constant as the system reaches the equilibrium state. In Fig.18.1, we note that there is ripple in $H(t)$. This is an indication that IVR is constrained and not completely ergodic. Periodic trajectories can be the cause for this observation.

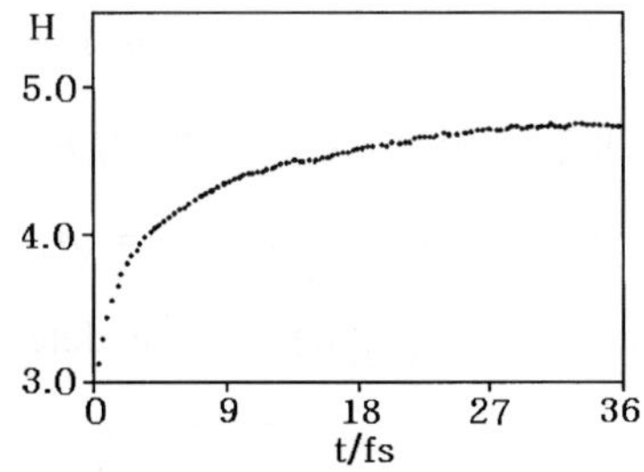

Fig.18.1 H function of H_2O

In Fig.18.1, $H(t)/t$ is the slope of the line joining the initial point and (t, $H(t)$). If $H(t)$ is steep, the average of $H(t)/t$ will also be large. This is especially obvious in the first 20 fs. Hence, the averaged $H(t)/t$ (denoted as k) can be a parameter describing the IVR rate. In the long term, it is found that k is right for describing the dynamical characters of a system. In the following, we will employ k to study the resonances in IVR of H_2O and DCN.

Fig.18.2(a) is the relation of k and n_s/n_t of H_2O with $n_t = 10$, $P(\equiv n_s + n_t + n_b/2) = 39$ and n_s is the variable. The highly excited states with $P = 39$ may not be so realistic. However, this will avoid the interference by 1:2 resonance as we study the 1:1 resonance, since then $n_b/2$ is much different from n_s and n_t. In the Figure, 1:1 resonance is very eminent. Fig.18.2(b) is the relation of k and $(n_b/2)/n_t$ of H_2O (with $n_t = 10$, $P = 39$, variable is n_b). From the Figure, 1:2 resonance is very obvious.

Fig.18.2(c) is the relation of k and n_s/n_t for DCN with $n_t = 6$, n_s as a variable. In the Figure, there are two peaks, corresponding to 1:1 and 2:3 resonances.

(a)

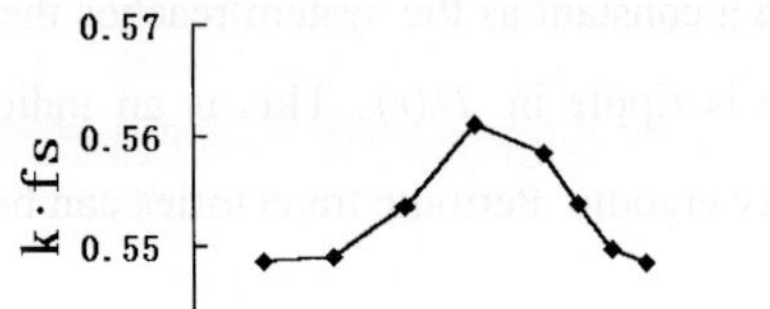

(b)

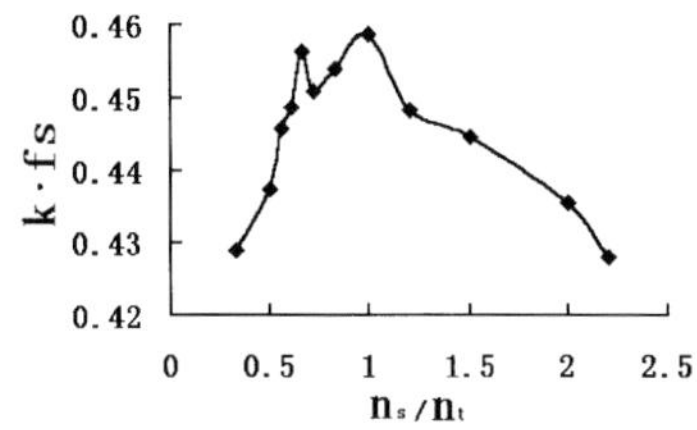

(c)

Fig.18.2 The relations of k and n_s/n_t (a), k and $(n_b/2)/n_t$ (b) for H_2O and k and n_s/n_t (c) for DCN

From Fig.18.2, it is seen that though the coupling coefficients of resonances 1:2 and 2:3 are smaller than that of 1:1, their strengths are stronger (In Fig.18.2, with sharper peaks). The cause is that a^+ and a (creation and destruction operators) are proportional to $(action)^{1/2}$. Hence, 1:2 and 2:3 resonances are proportional to $(action)^{3/2}$ and $(action)^{5/2}$, respectively. At high excitation, action is large and the strengths of these two resonances are larger than that of 1:1 resonance, which is proportional to action. The resonance peaks are about 2% (The base line is the off-resonance k value). Though this value is very small, it is large enough for the resonances to show up.

In this chapter, by employing the algebraic coset algorithm, we have constructed an *H* function in analogy with the *H* function in thermodynamics. Through this *H* function, resonances in high vibrational excitation can be studied in a convenient way. Though only the cases of H_2O and DCN are considered, the potential application of this algorithm deserves attention, especially in IVR for highly excited vibration.

References

18.1 J. R. Dorfman, *An Introduction to Chaos in Nonequilibrium Statistical Mechanisms*, Cambridge University Press, Cambridge, 1999.

18.2 D. Zheng, P. Wang, G. Wu, *Chem. Phys. Lett.*, 352(2002)79.

Chapter 19
The Dixon Dip and its destruction

19.1 The Dixon dip

As is well known, a resonance is a mimic of the pendulum motion for which there are two dynamical realms separated by its *separatrix* and two associated stable and unstable fixed points (See Sections 9.1,9.2). This is a classical viewpoint. Semiclassically, a quantum level (state) can be viewed as a subspace in the dynamical phase space. Therefore, levels can be classified as those lying below and above the separatrix (corresponding to a resonance) in the energy scale. For a pendulum, its oscillating frequency is expected to be smaller as the motion is close to the separatrix due to the nonlinear effect. Quantum mechanically, this will be reflected in that the nearest level energy spacing for those levels around the separatrix will reach a minimum. This is the so-called Dixon dip (Refs.19.1-4). This concept is integrated by the quantal and classical analogues. We note that the Dixon dip appears not only at the levels sharing a well-defined polyad number (see next section) as stated in the literature, but also when the resonance is very seriously perturbed by other complicated interactions, like in the systems of Henon–Heiles and quartic potentials for which resonance is hardly well defined. However, for the molecular systems like H_2O and DCN with multiple resonances, the Dixon dip will be destroyed by the overlapping of resonances which, as conjectured by Chirikov, will lead to chaos. This will be further analyzed by an independent Lyapunov exponent analysis as shown below.

Our concern is to use the Dixon dip as a possible way to explore the dynamics close to the transitional or dissociative state (See Chapter 20).

Before going further, we will briefly review the concept of an approximate quantum number via diabatic correlation for a vibrational system where most of its quantum numbers are destroyed by complicated interactions.

19.2 Approximate quantum numbers via diabatic correlation

This section is a short review of the approximate quantum numbers and diabatic correlation and is a preparation for the later sections. For their details, one is referred to Chapter 15.

For a highly excited vibration, most of its constants of motion or quantum numbers will commonly be destroyed due to complicated interactions including multiple resonances. Fortunately, even for those systems of which all the constants of motion except energy are destroyed, the destruction is often not so serious allowing some *approximate* quantum numbers, which are the quantum numbers of the eigenstates when the resonances were turned off, can be retrieved by diabatic correlation. In other words, the well-defined quantum numbers of the zeroth-order Hamiltonian are still the approximate constants of motion for the states of the system with full resonances. The retrieval of these approximate constants of motion is best realized via *diabatic* correlation as the parameters for the resonances are turned off gradually so that the eigenstates are correlated to the zeroth-order states. Therefore, from these approximate quantum numbers, we can construct the *approximate* polyad number corresponding to each resonance. In this way, we can categorize the states in groups defined by the (approximate) polyad numbers.

19.3 Dixon dips in the systems of Henon–Heiles and quartic potentials

The Henon–Heiles system is a two-dimensional harmonic oscillator with perturbing potential of the form

$$(p_x^2 + p_y^2 + x^2 + y^2)/2 + \lambda x(y^2 - x^2/3)$$

(λ is chosen as 0.05 in our calculation.) For this system, it is straightforward to cast the Hamiltonian in the second quantized language via

$$p_\alpha = i(a_\alpha^+ - a_\alpha)/\sqrt{2}$$

$$\alpha = (a_\alpha^+ + a_\alpha)/\sqrt{2} \qquad (\alpha = x, y)$$

for which a_α^+ and a_α are the creation and destruction operators, respectively. The perturbing potential is very complicated in the second quantized operators. However, embedded therein, there is only one term that is of resonance type, i.e., $a_x^+ a_y a_y + h.c.$ for which

$$P = n_x + n_y / 2$$

is conserved if all other perturbations are turned off. Here, n_x and n_y are the actions on the x and y coordinates. P is therefore not well defined. For this system, it is simple to construct the matrix Hamiltonian by employing the bases $|n_x, n_y>$ with n_x, n_y ranging from 0 to n which is chosen by arbitration. The eigenenergies are then obtained as the matrix Hamiltonian is diagonalized. This algebraic algorithm is also employed in our work for the systems of quartic potential, H_2O as well as DCN as demonstrated below. By reducing λ to zero, we can diabatically assign the quantum numbers of the zeroth-order states to those under the full perturbing potential. Hence, each state can be assigned with an approximate constant of motion P. Fig.19.1(a) shows the nearest level energy spacings for the states possessing a common P up to 7. The dips are obvious. (For clarity, only some of the cases with various P are shown. The numbers on the coordinate are the numbering of the nearest level energy spacings among the states. This is the same for other Figures.)

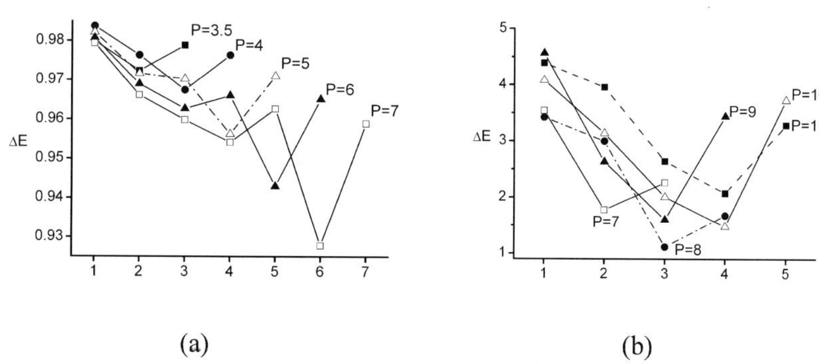

Fig.19.1 (a) The nearest level energy spacings for the states possessing a common P up to 7 for the system of Henon–Heiles potential. (For clarity, only some of the cases with various P are shown.) (b) The nearest level energy spacings among the states belonging to P = 7 to 11 for the system of quartic potential. The numbers on the coordinate are the numbering of the nearest level energy spacings among the states. This is the same for other Figures.

The quartic system has potential $\alpha x^2 y^2$ (α =0.25 in our calculation.) The Hamiltonian is

$$(p_x^2 + p_y^2)/2 + \alpha x^2 y^2$$

By second quantization, it can be seen that embedded in the quartic potential there is only one resonance, the Darling–Dennison (2:2) resonance :

$$a_x^+ a_x^+ a_y a_y + h.c.$$

Analogously, each eigenstate can be assigned with an approximate constant of motion

$$P = n_x + n_y$$

Shown in Fig.19.1(b) are the nearest-level energy spacings among the states belonging to $P = 7$ to 11. The dips are distinct.

In summary, it should be stressed that these results are not trivial if one notes that P is only approximate under the perturbing potentials.

19.4. Destruction of the Dixon dip under multiple resonances

A. H_2O system

For the system of H_2O, we consider the two resonances. One is between the two equivalent O–H stretches (with subscripts *s* and *t*) which is 1:1 resonance. The other is the Fermi resonance between the bend (with subscript *b*) and the two stretches. In second quantization, they are:

$$a_s^+ a_t + h.c.$$

and

$$a_s^+ a_b a_b + a_t^+ a_b a_b + h.c.$$

For this molecular system, the vibrational modes are represented as the Morse oscillator in the action form like $\omega(n+1/2) + X(n+1/2)^2$ up to the second anharmonicity. (For details, see Chapter 7) Four approximate polyad numbers can be defined for this system. They are $P_1 = n_s + n_t, P_2 = |n_s - n_t|$, $P_3 = n_s + n_b/2$ (this is the same as $n_t + n_b/2$) and $P_4 = n_s + n_t + n_b/2$ with n_s, n_t and n_b the actions on the stretches and bend. P_1 is due to the 1:1 resonance. P_2 is due to the linear combination of $a_s^+ a_b a_b + a_t^+ a_b a_b + h.c.$. P_3 is due to the Fermi resonance. P_4 is due to the 1:1 and Fermi resonances. P_4 is strict under both resonances.

For the case of P_1, the dips are not so sharp. This can be due to the 1:1 resonance not being so strong. Shown in Fig.19.2(a) is the case of $P_1 = 11$. Note that due to the equivalence of the two O–H stretches, there are almost degenerate states. They are the local modes in which the coupling between the two O–H stretches is small so that they behave more or less independently. This results in zero energy spacing in Fig.19.2(a). For the case of P_2, multiple dips are apparent. This is shown in Fig.19.2(b) for $P_2 = 2$. This can be attributed to the more complicated Fermi resonance. (It contains two terms and their hermitian conjugates: $a_s^+ a_b a_b + a_t^+ a_b a_b + h.c.$) For the case of P_3, dips appear in certain situations with smaller P_3. This is shown in Fig.19.2(c) for $P_3 = 3$. We note that due to multiple resonances, P_1, P_2 and P_3 are not well defined. However, the existence of the Dixon dips is still obvious in some cases. However, the destruction of the Dixon dip is also obvious due to multiple resonances. This will be further explored in the DCN system in the next section. P_4 is strictly conserved under 1:1 and Fermi resonances. However, the coexistence of these two resonances can cause the dip appearance in a complicated manner. The case of $P_4 = 6$ is shown in Fig.19.2(d). Indeed, there is an obvious concave structure in the energy spacings (It may be also viewed as two dips). For the case with larger P_4, the nearest level energy spacings show two distinct dips with zigzag structure (not shown). Apparently, this is due to the coexistence of 1:1 and Fermi resonances.

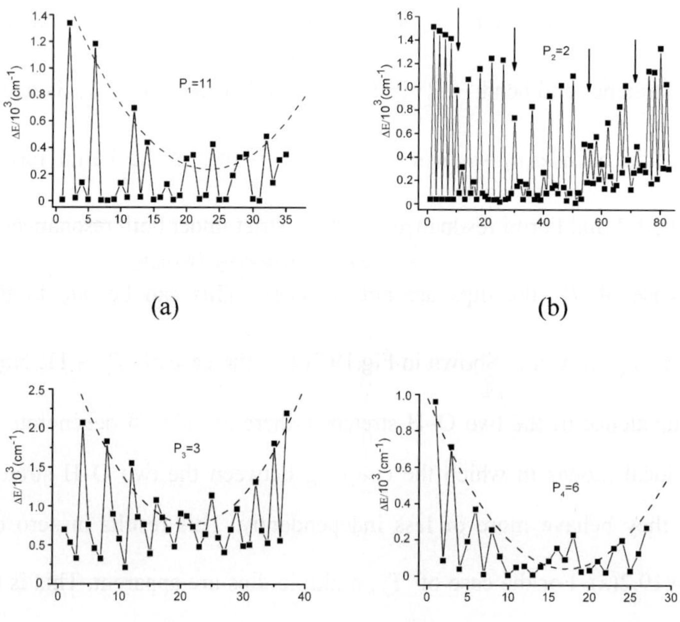

Fig.19.2 The nearest level energy spacings for the states of H_2O possessing (a) $P_1 = n_s + n_t =$ 11; (b) $P_2 = |n_s - n_t| = 2$; (c) $P_3 = n_s + n_b/2 = 3$; (d) $P_4 = n_s + n_t + n_b/2 =$ 6. Dashed lines and arrows are drawn for viewing the dips.

B. DCN system: overlapping of resonances and chaos

We consider the DCN system as composed of two Morse stretches with 1:1 and 2:3 resonances, due to the stretch frequency ratio of D–C to C–N (the frequencies are around 2681 cm^{-1} and 1949 cm^{-1}, respectively) being around 1.38, which is between 1 and 1.5. For this system, the approximate polyad numbers are:

$$P_1 = n_s + n_t$$

and

$$P_2 = n_s/2 + n_t/3$$

n_s and n_t are the actions on the D–C and C–N stretches, respectively. Shown in Fig.19.3 are the levels up to 45,000 cm^{-1} , which is close to dissociation and their unzipping by P_1 and P_2. (The calculation is based on the Morse model for the two stretches with 1:1 and 2:3 resonances. For details, see Chapter 14.

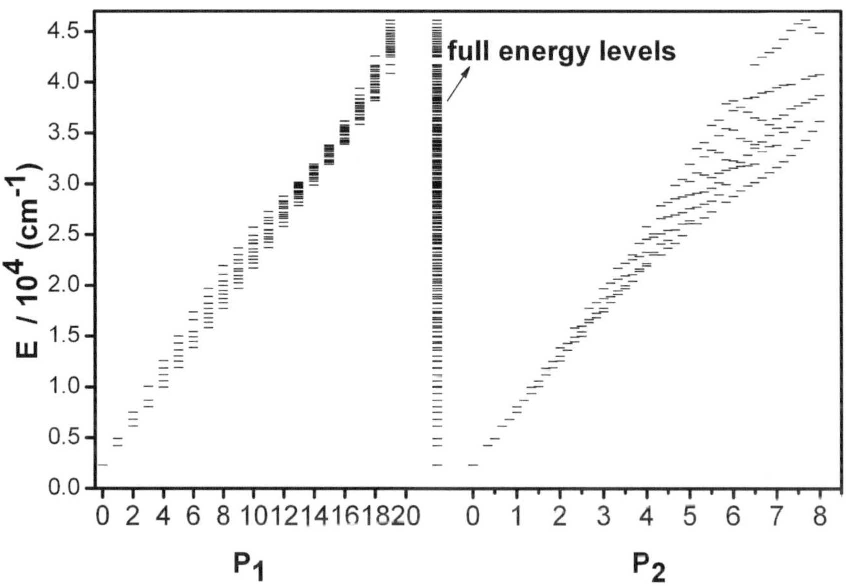

Fig.19.3 The levels of the two stretches of DCN up to 45 000 cm^{-1} and their unzipping by $P_1 = n_s + n_t$ and $P_2 = n_s/2 + n_t/3$.

For $P_1 < 7$ (i.e., for those levels < 20,000 cm^{-1}), the nearest level energy spacings show simple dips. (See Fig.19.4(a)). For higher levels (associated with $P_1 > 7$), the dip phenomenon is no longer evident and the nearest level energy spacings show a zigzag structure. This means that the 1:1 resonance is seriously perturbed by the 2:3 resonance. For $P_2 < 5$ (levels < 30 000 cm^{-1}) the dip phenomenon is not evident, while for larger P_2 (with level energy up to 45 000 cm^{-1}), the dips are distinct again. This is shown in Fig.19.4(b). By this comparison, we may conclude that for those levels in between 20,000 cm^{-1} and 30,000 cm^{-1}, the resonances are seriously mutually perturbed and chaos can predominate. For the lower and higher levels, only 1:1 or 2:3 resonance is operative and levels are more regular, dynamically.

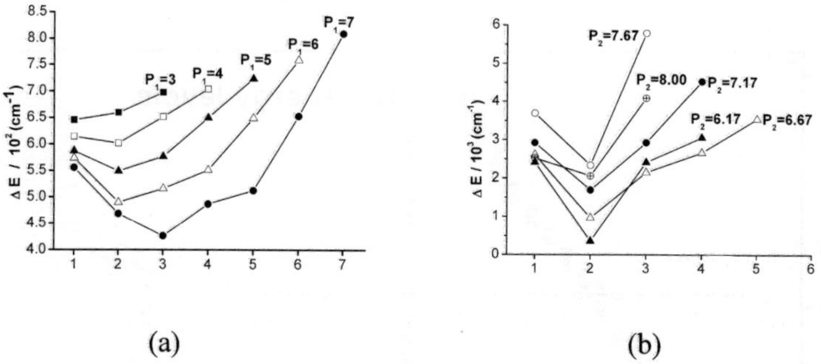

Fig.19.4 The nearest level energy spacings of DCN for the states with (a) $P_1 < 7$ (b) $P_2 > 5$

For treating the dynamics associated with a level (state), we may project the algebraic Hamiltonian in the coset space.

This treatment offers us the concept of chaos. For our purpose, only the averaged largest Lyapunov exponent is required, which is calculated by 200 randomly chosen initial points in the dynamical space corresponding to a state. (For details, see Chapter 13). The result is shown in Fig.19.5. Clearly, as the level is above 10,000 cm^{-1}, its dynamics is becoming chaotic. In this energy region, the dynamics is mainly by the 1:1 resonance, perturbed slightly by the 2:3 resonance. The dynamics reaches the highest degree of chaoticity as level energy reaches 25,000 cm^{-1}. In between 20,000 cm^{-1} and 30,000 cm^{-1}, both 1:1 and 2:3 resonances are operative. Their overlapping, as conjectured by Chirikov, leads to chaos. For the levels above 30,000 cm^{-1}, the diminishing of the 1:1 resonance leaves only the 2:3 resonance prominent. Meanwhile, the overlapping of the two resonances and the Lyapunov exponent are also diminishing. Hence, the degree of chaoticity drops slightly even when the level energy increases more!

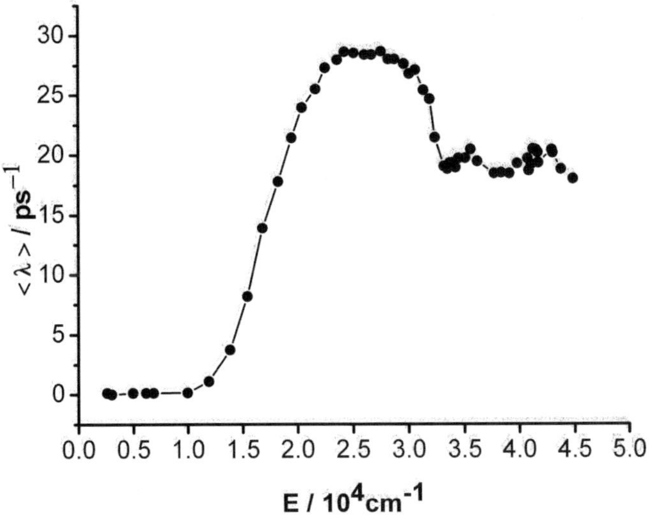

Fig.19.5 The averaged largest Lyapunov exponent $<\lambda>$ as a function of the level energy. See text for its detail calculation.

In summary, our work shows that this dip phenomenon is so evident even for a resonance that is seriously perturbed by other interactions like in the systems of Henon–Heiles and quartic potentials for which the polyad numbers are not well-defined and only approximate. This demonstrates a *global* viewpoint for unfolding the dynamics which does not rely on individual levels. In general, the dip phenomenon is distinct for a simple resonance. The emerging of multiple resonances definitely will complicate the level structure and may cause the dip phenomenon to be less obvious. We demonstrated this for the system of H_2O. For the system of DCN, the dips are only apparent for the low and high levels while not for those levels in between. In terms of Chirikov's conjecture, this is due to the overlapping of resonances which will lead to chaos. Independently, this issue is analyzed in terms of the degree of chaoticity by the averaged Lyapunov exponent. These two results show consistent interpretations. This algorithm can be of potential value in the elucidation of the dynamical properties of the highly excited vibrational system, including the *transitional state*, of which multiple resonances, and therefore chaos, is prevailing.

However, only a little is known about its dynamics and at its very primitive stage.

References

19. 1 R.N.Dixon, *Trans. Farad. Soc.* 60 (1964) 1363

19. 2 M.S.Child, *J. Mol. Spectrosc.* 210 (2001) 157

19. 3 S.Yang, V.Tyng and M.E.Kellman, *J. Phys. Chem. A* 107 (2003) 8345

19. 4 J.Svitak, Z.Li, J.Rose and M.E.Kellman, *J. Chem. Phys.* 102 (1995) 4340

19.5 H. Wang, P. Wang and G. Wu, *Chem.Phys.Lett.*, 399 (2004) 78

Chapter 20
Chaos in transitional states

20.1 Chaos in dissociation

The basic characteristic of a nonlinear system is that its classical frequency depends on its action. The classical frequency ω_c can be written as:

$$\partial E / \partial n$$

here, *E* is energy and *n* is action. For an harmonic oscillator, ω_c is a constant and the system is linear. For a Morse oscillator, ω_c is smaller as *n* is larger. In this formula, if $\partial n = 1$, then ∂E is the energy spacing between two neighboring levels. At dissociation, ω_c is close to 0. This is because that the nearest level energy spacing is zero at dissociation (See Section 1.2). The quantum number (action) n_0 at dissociation is

$$n_0 = \frac{-\omega}{2X} - 1$$

The frequency at dissociation, ω_c^0 is $-X$. Since $-X << \omega$, ω_c^0 is very small.

The above physics is very similar to the pendulum motion. At low excitation. the motion is around the stable fixed point with small amplitude. At high excitation close to dissociation, the motion is just like that of the pendulum around the unstable fixed point. In Chapter 9, we have addressed this concept and pointed out that chaos appears around the unstable fixed point. Therefore, we conclude that *dissociation, or more generally, a transitional state is accompanied by chaos.*

The structure and characteristics of this chaos are not well known for the moment. This is a topic relating to the very basic physics and chemistry. In this chapter, we will try to address the related topics. Our approach is just a premature trial.

20.2 Chaos in the transitional states of bend motion

A similar phenomenon appears in the high excitation of bend in intramolecular rotation. For example, in acetylene, the H atom of the C–H bend at high excitation will overcome the potential barrier to migrate to another C atom to form vinylidene. The top of the barrier corresponds to the unstable fixed point of the pendulum motion. In HCN, the highly excited bend motion of C–H may cause the H atom to overcome the energy barrier and to migrate to the N atom to form the stable HNC. Another example is the migration of the H atom to form linear HPC by the high excitation of the C–H bend motion in HCP. HPC is unstable with the H atom at the unstable saddle (fixed) point. At even higher excitation, the H atom will rotate around the C–P core. The case of acetylene is more complicated, in which the C–H bend and stretches of C–H and C–C will be in resonance (See Section 12.1). In its transitional state, the C–H bend frequency is reduced and the bend will be in multiple resonances with the stretch. This system involves two C–H bends and three stretches. So, the dynamics of its transitional state is very complicated. All these transitional states due to the highly excited C–H bend motion are unstable delocalized states. The physics is just similar to the unstable fixed point of the pendulum motion, of which the classical frequency of the C–H bend motion, ω_c, is close to 0 and there will be associated chaotic motion. The topic of the unique character of this chaotic motion is of interest. The coupling of this motion with other vibrational modes can be the crucial factor.

More specifically, take the HCN case as an example. The main coupling is between the C–H stretch and C–H bend. The stretch frequency ω_s is 3311 cm^{-1} and for the bend, ω_b is 713 cm^{-1}. (The coupling between the C–H bend and C–N stretch is very small.) By energy consideration, the coupling orders are 1:4 (energy difference 459 cm^{-1}) and 1:5 (energy difference 254 cm^{-1}). In HNC, the frequency of the N–H stretch, ω_s, is 3653 cm^{-1} and that of the bend, ω_b, is 477 cm^{-1}. Hence, the couplings can be of the orders 1:7 (energy difference 314 cm^{-1}) and 1:8 (energy difference 163 cm^{-1}). In the transitional state, ω_c of the H bend around C–N will be less than 400

cm^{-1}. The bend will be in multiple resonances with the stretch which is between 3311 cm^{-1} and 3653 cm^{-1}. (The stretch is at low excitation with a very small nonlinear effect.) The case of HCP is similar. ω_b of the C–H bend is 698 cm^{-1} which can be in 1:2 resonance with the C–P stretch of which ω_s is 1301 cm^{-1}. As the energy of the C–H bend is high enough for the system to be of the linear saddle structure, CPH, the bend frequency will be reduced so that multiple resonances with the C–P stretch are possible.

Here, we give an estimation of the multiple resonances by simple evaluation. Suppose the stretch has $\omega_s = 3000\,cm^{-1}$, $X_{ss} = -100\ cm^{-1}$ and the bend has $\omega_r = 1500\,cm^{-1}$, $X_{rr} = -10\ cm^{-1}$. The bend quantum number n_r is 0 to 74, which is the quantum number at dissociation as a Morse oscillator. We define the resonance order as ω_s^c / ω_r^c with $\omega_s^c = \omega_s + 2X_{ss}(n_s + 1/2)$ and $\omega_r^c = \omega_r + 2X_{rr}(n_r + 1/2)$. Listed in Table 20.1 are the resonance orders as energy or n_r increases with $n_s = 0$. In evaluating resonance order, the energy width (uncertainty) is taken as 100 cm^{-1}. The observation is that in the levels within 1000 cm^{-1} below dissociation, overlaps of resonances are possible.

Table 20.1 The resonance order as a function of energy and n_r

Energy (cm^{-1})	n_r	Resonance order
2223	0	} 1:2
34203	26	} 1:2
44540	38	} 1:3
49022	45	} 1:4
51723	50	} 1:5
53523	54	} 1:6
54302	56	} 1:7
55002	58	} 1:8
55622	60	} 1:9
55902	61	} 1:10
56162	62	} 1:11

56403	63	} 1:12
56403	63	} 1:13, 1:14
56622	64	} 1:14, 1:15
56823	65	} 1:16, 1:17
57002	66	} 1:18, 1:19
57163	67	} 1:19--1:22
57302	68	} 1:23--1:28
57423	69	} 1:25--1:35
57522	70	} 1:30--1:53
57602	71	} 1:34--1:96
57663	72	} 1:34--1:290
57723	74	

In summary, we know that transitional state induced by bend motion is unstable. It will involve multiple resonances with the stretches due to the frequency reduction of the bend motion. Chaos in the transitional state is certainly related to the conjecture by Chirikov that resonance overlap will lead to chaos. (See Section 9.2)

20.3 HCN, HNC and the delocalized state

HCN、HNC and the delocalized (transitional) state form a typical system induced by the bend motion of the H atom. Experimentally, the data for its very high excitation are scarce. For the moment of our probe, the vibrational states by *ab initio* quantal calculation are adopted. We consider the system as composed of three Morse oscillators: the stretch of H against C–N, the bend of H and the stretch of C–N. Their coordinates are labeled by subscripts: 1, 2 and 3, respectively. The Hamiltonian is

$$\omega_1(n_1+1/2)+X_{11}(n_1+1/2)^2+\omega_2(n_2+1)+X_{22}(n_2+1)^2$$
$$+\omega_3(n_3+1/2)+X_{33}(n_3+1/2)^2$$

together with the anharmonic terms:

$$X_{12}(n_1+1/2)(n_2+1)+X_{13}(n_1+1/2)(n_3+1/2)+X_{23}(n_2+1)(n_3+1/2)$$

Here, ω and X denote the harmonic and anharmonic coefficients (in the unit of cm^{-1}). n is the quantum number or action.

The energy transfer or couplings among the modes are:

$$\sum K_{k_1k_2k_3}O_1^{|k_1|}O_2^{|k_2|}O_3^{|k_3|}$$

The coefficient K is the strength. k_1, k_2, k_3 are such that $k_1\omega_1 + k_2\omega_2 + k_3\omega_3$ has to be smaller and $|k_1|+|k_2|+|k_3|$ (the coupling order) can not be too large. As $k_i > 0$, O_i is a_i^+; as $k_i < 0$, O_i is a_i. a_i^+ and a_i are the creation and destruction operators, respectively.

This is an algebraic Hamiltonian. With appropriate basis space, $|n_1\rangle|n_2\rangle|n_3\rangle$ – for the delocalized state, n_1, n_3 are chosen as 0, 1, 2 and n_2 is 40–70. This is because that delocalized state is mainly due to the high excitation of the H bend – and a set of proposed ω, X and K, we can construct the corresponding Hamiltonian matrix by which the eigenenergies can be obtained. By fitting the calculated eigenenergies to the quantal levels (Refs. 20.1,20.2), it is possible to finalize ω, X, K. We list our results in Table 20.2 in which the data of HCN and CNH are separated into two groups. The consideration is that for higher levels, the coefficients ω, X. K may vary. (Indeed, calculation confirms this.) The average fit error is 10 cm^{-1}. For the delocalized state, it is 5 cm^{-1}. Different conditions for fit convergence may cause uncertainty. This is more evident for X and K. Their uncertainty is about 10% (For the delocalized state, it is about 5%) .

For the levels of HCN, from 713 cm^{-1} to 12,328 cm^{-1} and from 12,364 cm^{-1} to 23,047 cm^{-1}, ω and X are roughly the same as shown in the Table. The main difference is that low excitation is with weak resonance. At high excitation, K_{11-2}, K_{1-3-1}, K_{1-2-1}, K_{04-1} are prominent. Indeed, multiple resonances between the stretches and bend are obvious.

For HNC, the low energy range is from 5023 cm^{-1} to 12,703 cm^{-1} and the high energy range is from 12,708 cm^{-1} to 15,505 cm^{-1}. Mode frequencies in these two ranges are similar. The most significant is that the bend frequency drops from 728 cm^{-1}, 736 cm^{-1} to 492 cm^{-1} and 498 cm^{-1}. For the low energy range, K_{10-2}, K_{11-2}; K_{1-2-1}, K_{1-3-1} are prominent. For the high energy range, except where these couplings are strong, K_{03-1}, K_{04-1} are also not small. All these show the multi-resonances.

The energy range for the delocalized state is 15,205–18,346cm^{-1}. ω_2 is significantly small, only 258 cm^{-1}. Since n_2 is very large (40–70), the fitted K's are small. However, this does not mean that resonance is small. Instead, they are still strong. K_{1k_2-1}(k_2 = -2,-3,-4); K_{12-2}, K_{13-2} are of similar orders. Multiple resonances are evident.

Table 20.2 The fitted ω, X, K (cm^{-1}) of HCN、HNC and delocalized state (level energy is with respect to the zero point energy of HCN)

	HCN		HNC		delocalized state
	713-12328	12364-23047	5023-12703	12708-15505	15206-18346
ω_1	3442.1	3494.6	3284.1	3317.1	3342.22
ω_2	728.2	736.4	492.2	498.0	258.35
ω_3	2129.9	2109.4	2069.4	2046.0	2227.00
X_{11}	-51.4	-60.8	-69.4	-86.8	-91.66
X_{22}	-2.6	-3.7	-26.0	-5.3	0.11
X_{33}	-10.7	-7.3	-28.7	-29.1	-5.32
X_{12}	-19.1	-25.2	18.1	-32.7	-23.19
X_{13}	-15.9	-15.3	-93.5	-84.7	-27.41
X_{23}	-3.1	-3.7	-62.4	-36.9	-23.59
K_{03-1}			-9.3	-6.3	
K_{04-1}		-2.1		-5.8	
K_{10-2}	-2.9		-58.2	-46.8	
K_{11-2}		9.1	21.7	8.4	
K_{1-2-1}		3.2	-76.5	-73.9	-0.45

$K_{1\text{-}3\text{-}1}$	-5.6	13.6	8.7	-0.79
$K_{1\text{-}4\text{-}1}$				-0.13
$K_{12\text{-}2}$				1.09
$K_{13\text{-}2}$				0.17
$K_{14\text{-}2}$				-0.04

20.4 The Lyapuov exponent for transitional chaos

For calculating the Lyapunov exponents of the states near transition and dissociation, the following two formulae are drawn:

A. A pendulum coupled with an harmonic oscillator

The pendulum is a mimic of the bend. The Hamiltonian and Hamilton's equations of motion are:

$$v^2/2 - K\cos\phi + p_x^2/2 + \omega^2 x^2/2 + kp_x v$$

and

$$\begin{aligned} d\phi/dt &= v + kp_x \\ dv/dt &= -K\sin\phi \\ dx/dt &= p_x + kv \\ dp_x/dt &= -\omega^2 x \end{aligned}$$

B. For a Morse oscillator coupled with an harmonic oscillator

The Hamiltonian and Hamilton's equations of motion are:

$$v^2/2 + D[1-\exp(-ar)]^2 + p_x^2/2 + \omega^2 x^2/2 + kp_x v$$

and

$$\begin{aligned} dr/dt &= v + kp_x \\ dv/dt &= -2aD[\exp(-ar) - \exp(-2ar)] \\ dx/dt &= p_x + kv \\ dp_x/dt &= -\omega^2 x \end{aligned}$$

In the above expressions, v is the angular momentum of the bend or the momentum of the Morse oscillator. p_x is that for the harmonic oscillator. D and a are the parameters for the Morse oscillator. ω is the angular frequency of the harmonic oscillator. K and k are parameters for the pendulum potential and the momentum coupling strength.

For both cases, we note that close to the bend transitional and dissociative states, the momentum of bend or Morse oscillator v is close to 0 and the motion is *defined* by p_x. Since p_x of the harmonic oscillator is of high frequency, the motion of the bend as its transitional state is approached is inevitably wiggling. This is also true as a Morse oscillator is close to the dissociation.

Shown in Fig.20.1 is the plot of the Lyapunov exponent, λ, against the initial v with initial $(\phi, v, x, p_x)_0 = (0, v, 1.1, 1.1)$ for the Case A with parameters: $K = 1.1, \omega = 4.5, k = -0.15$.

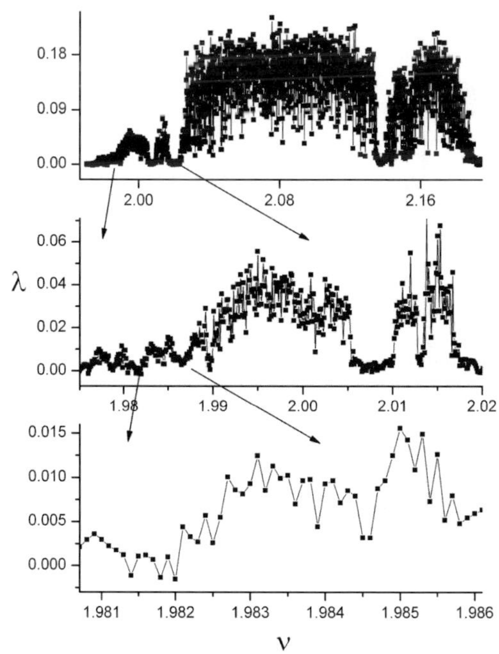

Fig.20.1 The plot of the Lyapunov exponent (λ) against the initial v with initial $(\phi, v, x, p_x)_0 = (0, v, 1.1, 1.1)$ for Case A and parameters: $K = 1.1, \omega = 4.5, k = -0.15$. The self-similarity structure is apparent.

Under these parameters below the transition (which corresponds to $v = 1.98$), all trajectories are regular. As the system energy is just above the transitional point, chaos starts appearing.

Apparently, as shown, there is *structure* in the Lyapunov exponent distribution. Interestingly, there is evidence of self-similarity. This is demonstrated in the Figure also.

For the system of a Morse oscillator coupled with an harmonic oscillator, even far below the dissociation, the Lyapunov exponent is nonzero. This means that chaos starts emerging before the dissociation. This is different from the pendulum case. Shown in Fig.20.2 is the plot of the Lyapunov exponent, λ, against the initial v. Other initial conditions are: D = 85, a = 1.1, k = -0.15 and $(r,v,x,p_x)_0 = (0.1,v,0.1,7.0)$ Fig.20.3 shows the corresponding levels. In Fig.20.2, it is interesting to note that before the steady increase of the Lyapunov exponent, there is *structure* in its distribution.

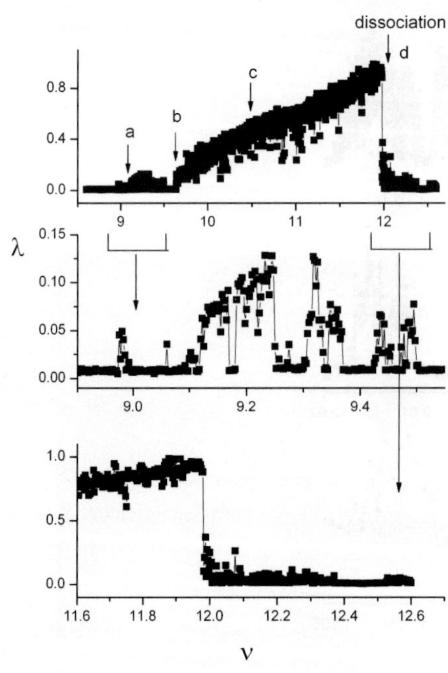

Fig.20.2 The plot of the Lyapunov exponent against the initial v for Case B. Other initial conditions are: D = 85, a = 1.1, k = -0.15 and $(r,v,x,p_x)_0 = (0.1,v,0.1,7.0)$. Note that before the steady increase of the Lyapunov exponent, there is structure in its distribution. See text for details.

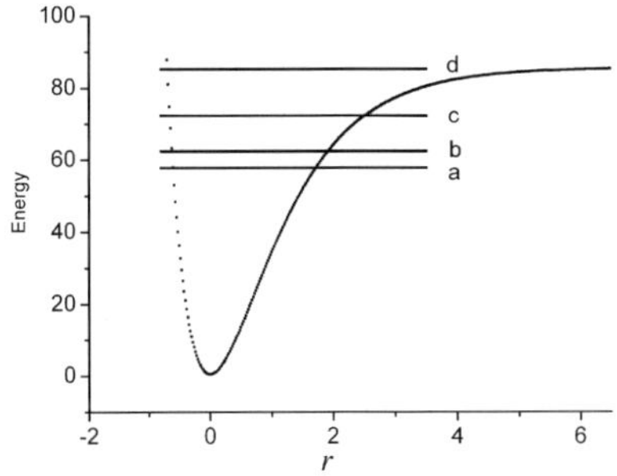

Fig.20.3 The levels as shown in Fig.20.2

The work up to this point is very premature. Nevertheless, I would like to show these very crude observations. Indeed, it demonstrates a foreseeable fruitful dynamical essence in the *transitional chaos.* This is a field that deserves our attention. Hopefully in the near future, more about its physics will be unfolded. This is also my intention.

References

20. 1 J M Bowman, B Gazdy, J A Bentley, T J Lee and C E Dateo, *J. Chem. Phys.* 99 (1993)308

20. 2 J A Bentley, C M Huang and R E Wyatt, *J. Chem. Phys.* 98 (1993)5207

Index